ROTAR

The Auxiliaries

Unit I, Lesson 9
Third Edition

▼
▼
▼

by Kate Van Dyke

Published by

PETROLEUM EXTENSION SERVICE
Continuing & Innovative Education
The University of Texas at Austin
Austin, Texas

in cooperation with

INTERNATIONAL ASSOCIATION
OF DRILLING CONTRACTORS
Houston, Texas

1999

Library of Congress Cataloging-in-Publication Data

Van Dyke, Kate, 1951—
The auxillaries / written by Kate Van Dyke. — 3rd ed.
p. cm. — (Rotary drilling series ; unit I, lesson 9)
ISBN 0-88698-184-0 (pbk. : alk. paper)
1. Oil well drilling—Equipment and supplies. I. University of Texas at Austin. Petroleum Extension Service. II. Title.
III. Series: Rotary drilling series; unit 1, lesson 9.
TN871.2.A79 1999
622'.3381—dc21 98-24789
CIP

First Edition published 1967. Third Edition 1999.
Fifth Impression 2011
Printed in the United States of America

Catalog no. 2.10930
ISBN 0-88698-184-0

Contents

Figures

Foreword

For many years, the Rotary Drilling Series has oriented new personnel and further assisted experienced hands in the rotary drilling industry. As the industry changes, so must the manuals in this series reflect those changes.

The revisions to both text and illustrations are extensive. In addition, the layout has been "modernized" to make the information easy to get; the study questions have been rewritten; and each major section has been summarized to provide a handy comprehensive check for the student.

Ron Baker

Acknowledgments

The extensive reorganization and rewriting of this book would not have been possible without the help of Tom Thomas of Sedco Forex, who provided many of the contacts for obtaining information. Thanks to David Reid at Varco for sending his article on the history of pipe handling, and Steve Legler at Varco for answering questions about kelly spinners and power slips. Randy Leo at M/D Totco was ever cheerful while explaining instrumentation. Margie Griffiths at Aqua-Chem's wonderful web site and Paul Choules at MECO sent much helpful information about watermakers. Thanks to Alison Kelman at Brandt for providing manuals and brochures about cleanup equipment all the way from Scotland. Finally, thanks to Paul Neely of Task Force Tips for opening up the world of nozzles and monitors, and to Stephen Wright of Sedco Forex for an eye-opening experience at Texas A&M's firefighting school. Both were generous with their time and expertise

In spite of the assistance PETEX received in writing and illustrating this book, PETEX is solely responsible for its contents. Also, while every effort was made to ensure accuracy, this manual is intended only as a training aid. Nothing in it should be considered approval or disapproval of any specific product or practice.

PETEX wishes to thank industry reviewers—and our readers—for invaluable assistance in the revision of the Rotary Drilling Series. On the PETEX staff, Kathryn Roberts saw production through from idea to book, Deborah Caples designed the layout, and Doris Dickey proofread innumerable versions.

Units of Measurement

Throughout the world, two systems of measurement dominate: the English system and the metric system. Today, the United States is almost the only country that employs the English system.

The English system uses the pound as the unit of weight, the foot as the unit of length, and the gallon as the unit of capacity. In the English system, for example, 1 foot equals 12 inches, 1 yard equals 36 inches, and 1 mile equals 5,280 feet or 1,760 yards.

The metric system uses the gram as the unit of weight, the metre as the unit of length, and the litre as the unit of capacity. In the metric system, for example, 1 metre equals 10 decimetres, 100 centimetres, or 1,000 millimetres. A kilometre equals 1,000 metres. The metric system, unlike the English system, uses a base of 10; thus, it is easy to convert from one unit to another. To convert from one unit to another in the English system, you must memorize or look up the values.

In the late 1970s, the Eleventh General Conference on Weights and Measures described and adopted the Système International (SI) d'Unités. Conference participants based the SI system on the metric system and designed it as an international standard of measurement.

The *Rotary Drilling Series* gives both English and SI units. And because the SI system employs the British spelling of many of the terms, the book follows those spelling rules as well. The unit of length, for example, is *metre*, not *meter*. (Note, however, that the unit of weight is *gram*, not *gramme*.)

To aid U.S. readers in making and understanding the conversion to the SI system, we include the following table.

English-Units-to-SI-Units Conversion Factors

Quantity or Property	English Units	Multiply English Units By	To Obtain These SI Units
Length, depth, or height	inches (in.)	25.4	millimetres (mm)
		2.54	centimetres (cm)
	feet (ft)	0.3048	metres (m)
	yards (yd)	0.9144	metres (m)
	miles (mi)	1609.344	metres (m)
		1.61	kilometres (km)
Hole and pipe diameters, bit size	inches (in.)	25.4	millimetres (mm)
Drilling rate	feet per hour (ft/h)	0.3048	metres per hour (m/h)
Weight on bit	pounds (lb)	0.445	decanewtons (dN)
Nozzle size	32nds of an inch	0.8	millimetres (mm)
Volume	barrels (bbl)	0.159	cubic metres (m^3)
		159	litres (L)
	gallons per stroke (gal/stroke)	0.00379	cubic metres per stroke (m^3/stroke)
	ounces (oz)	29.57	millilitres (mL)
	cubic inches ($in.^3$)	16.387	cubic centimetres (cm^3)
	cubic feet (ft^3)	28.3169	litres (L)
		0.0283	cubic metres (m^3)
	quarts (qt)	0.9464	litres (L)
	gallons (gal)	3.7854	litres (L)
	gallons (gal)	0.00379	cubic metres (m^3)
	pounds per barrel (lb/bbl)	2.895	kilograms per cubic metre (kg/m^3)
	barrels per ton (bbl/tn)	0.175	cubic metres per tonne (m^3/t)
Pump output and flow rate	gallons per minute (gpm)	0.00379	cubic metres per minute (m^3/min)
	gallons per hour (gph)	0.00379	cubic metres per hour (m^3/h)
	barrels per stroke (bbl/stroke)	0.159	cubic metres per stroke (m^3/stroke)
	barrels per minute (bbl/min)	0.159	cubic metres per minute (m^3/min)
Pressure	pounds per square inch (psi)	6.895	kilopascals (kPa)
		0.006895	megapascals (MPa)
Temperature	degrees Fahrenheit (°F)	$\frac{°F - 32}{1.8}$	degrees Celsius (°C)
Thermal gradient	1°F per 60 feet	—	1°C per 33 metres
Mass (weight)	ounces (oz)	28.35	grams (g)
	pounds (lb)	453.59	grams (g)
		0.4536	kilograms (kg)
	tons (tn)	0.9072	tonnes (t)
	pounds per foot (lb/ft)	1.488	kilograms per metre (kg/m)
Mud weight	pounds per gallon (ppg)	119.82	kilograms per cubic metre (kg/m^3)
	pounds per cubic foot (lb/ft^3)	16.0	kilograms per cubic metre (kg/m^3)
Pressure gradient	pounds per square inch per foot (psi/ft)	22.621	kilopascals per metre (kPa/m)
Funnel viscosity	seconds per quart (s/qt)	1.057	seconds per litre (s/L)
Yield point	pounds per 100 square feet (lb/100 ft^2)	0.48	pascals (Pa)
Gel strength	pounds per 100 square feet (lb/100 ft^2)	0.48	pascals (Pa)
Filter cake thickness	32nds of an inch	0.8	millimetres (mm)
Power	horsepower (hp)	0.75	kilowatts (kW)
Area	square inches ($in.^2$)	6.45	square centimetres (cm^2)
	square feet (ft^2)	0.0929	square metres (m^2)
	square yards (yd^2)	0.8361	square metres (m^2)
	square miles (mi^2)	2.59	square kilometres (km^2)
	acre (ac)	0.40	hectare (ha)
Drilling line wear	ton-miles (tn•mi)	14.317	megajoules (MJ)
		1.459	tonne-kilometres (t•km)
Torque	foot-pounds (ft•lb)	1.3558	newton metres (N•m)

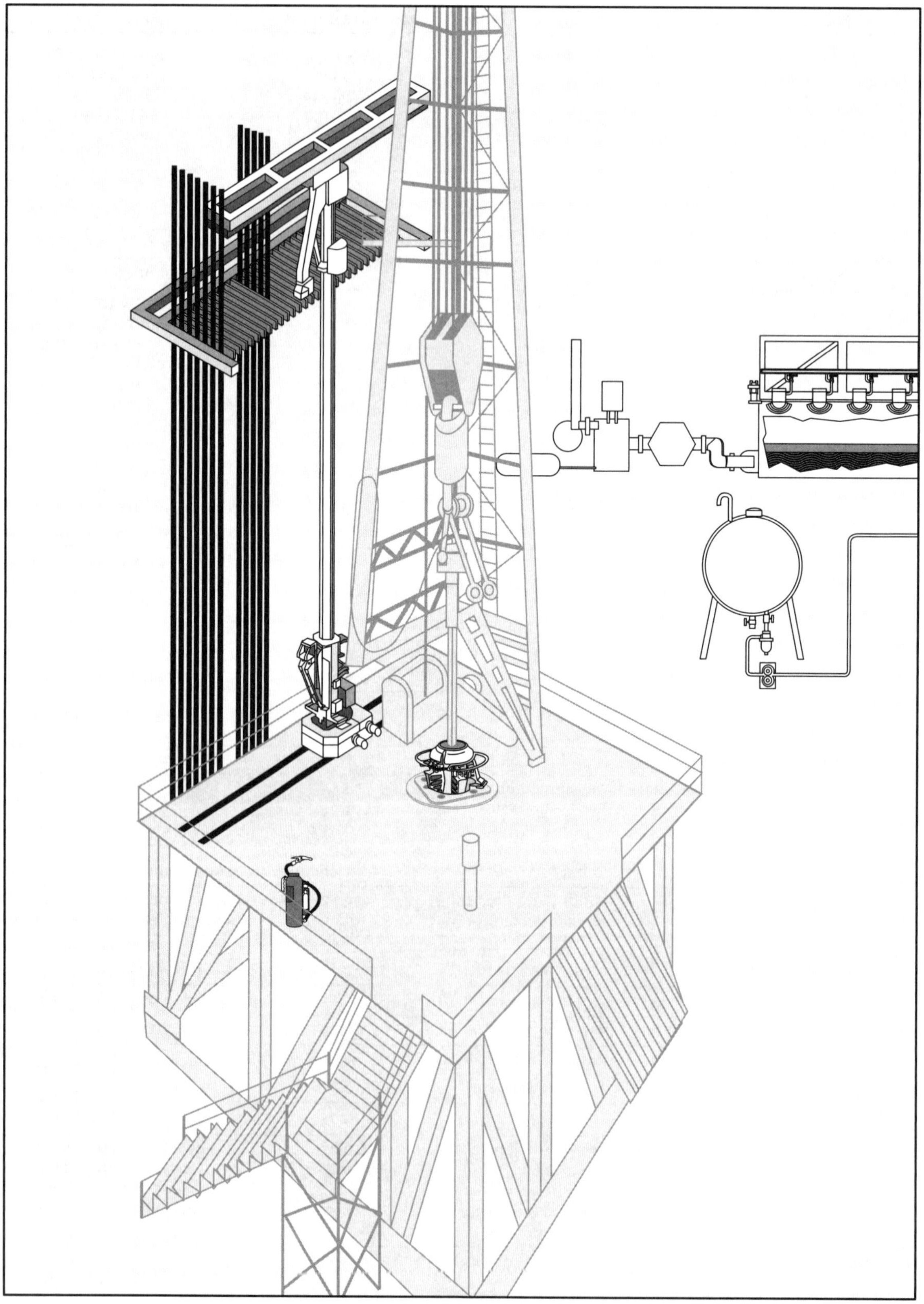

Introduction

While much of the equipment on a rig site is the huge machinery that does the main work of drilling, many other tools and pieces of equipment round out the typical drilling operation. Besides the drilling equipment are the tools that allow the crew to work with pipe, the instruments that monitor drilling, the equipment that provides water and electricity, and the equipment that makes the job safer.

The drilling industry continuously improves both primary and auxiliary equipment. In particular, recent increases in deep offshore drilling and in directional and horizontal drilling have motivated numerous changes. Advances in automation have been especially dramatic. Automation on the rig is the use of automatic mechanical or electronic devices to replace human observation, labor, and decisionmaking. Automation replaces manual, repetitive tasks with machines, which removes people from hazardous work and locations. It allows more precise control of processes, and produces more consistent quality. Moreover, it allows one person to control several functions simultaneously, and it can make a company or an industry more competitive by reducing costs and waste.

Other lessons in this series go into detail about many of the auxiliary tools used on a drilling rig. This book covers equipment that is not mentioned in other lessons or is mentioned only briefly.

Pipe-Handling Equipment

▼
▼
▼

Automation has changed, and in some cases eliminated, many manual tasks on the rig floor. On rigs that have automatic equipment, handling pipe and slips is no longer the heavy work it used to be. In many ways, automation has transformed the floorhand from a laborer to an operator. From the pipe handling side of the operation, automatic equipment includes kelly spinners, spring slips and power slips, and automated pipe-handling and racking systems.

Kelly Spinners

On rigs using a kelly-and-rotary-table system (instead of a top drive) to rotate the bit, kelly spinners are great labor saving devices. A *kelly spinner* (fig. 1) is a pneumatic (powered by compressed air)

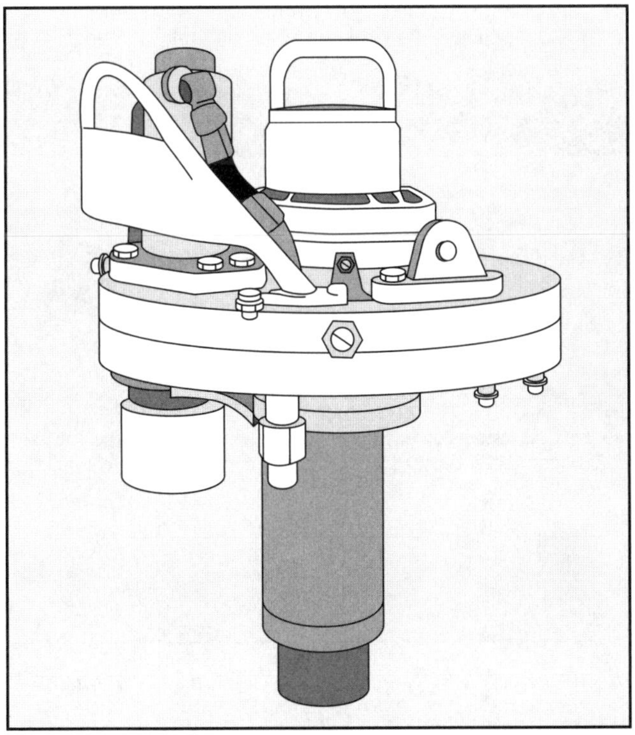

Figure 1. This pneumatic kelly spinner spins both directions.

or hydraulic (powered by a liquid called hydraulic fluid) motor attached to the top of the kelly or to the bottom of the swivel (fig. 2). The kelly spinner's job is to rapidly turn, or spin, the kelly, mainly when making a connection—that is, when adding a joint of drill pipe to the string after the kelly has been drilled down.

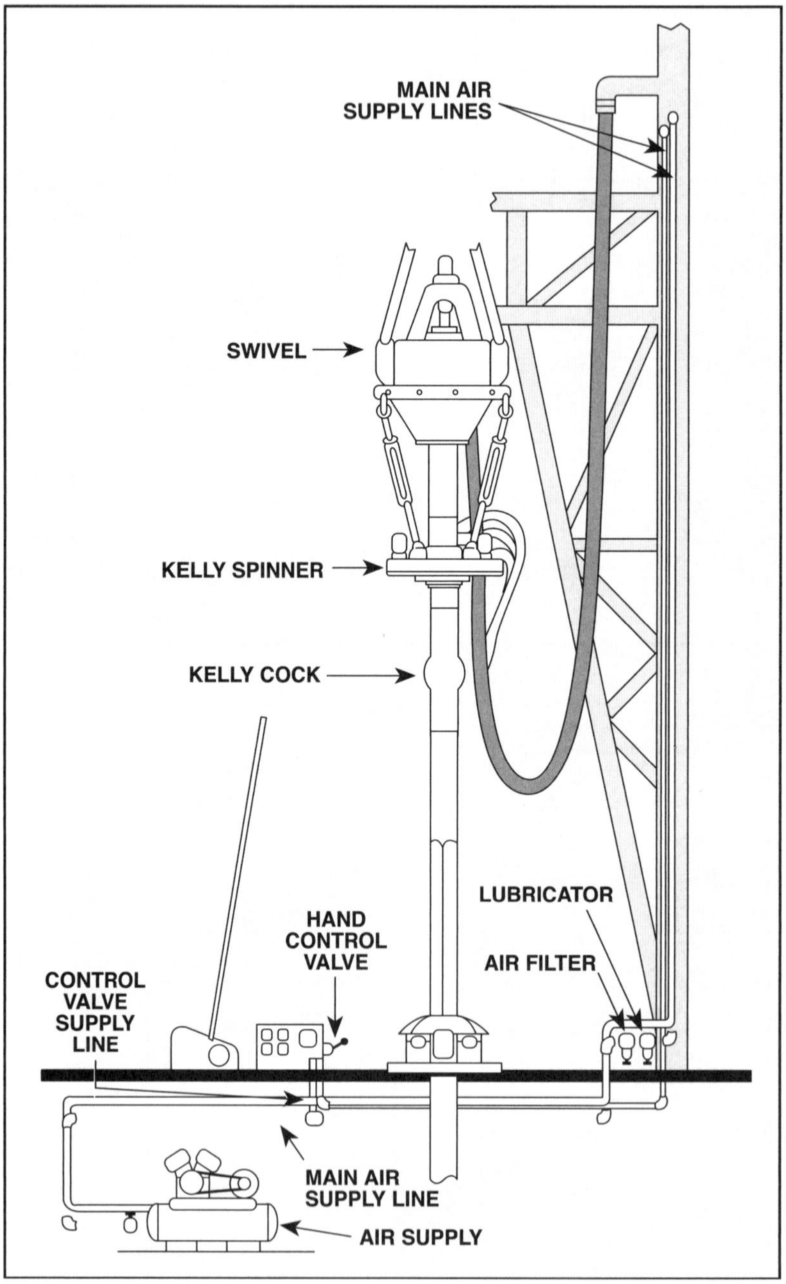

Figure 2. This air-powered kelly spinner has a hand control on the rig floor. Hydraulic models work the same way.

To make a connection on a conventional rotary rig, crew members first set the slips to suspend the drill string in the rotary table. They then use the tongs to break out, or loosen, the connection. Once the connection is broken, the driller rapidly turns the rotary table to the right to spin out the drill string from the kelly. The spinning motion unscrews the kelly threads from the drill pipe tool joint. The floorhands then swing the kelly over to the drill pipe joint in the mousehole and stab the kelly into the joint. They latch the backup tongs around the tool joint of the drill pipe in the mousehole. The driller then engages the kelly spinner to spin up the kelly into the drill pipe. At this point, the crew can either tong the kelly up to final tightness or, more likely, the driller can pick up the kelly and spun up joint and let the floorhands stab the new joint and attached kelly into the drill string hanging in the rotary. The driller engages the kelly spinner and spins up the kelly and joint into the drill string. The floorhands then tong up the drill pipe to final tightness, the driller lowers the drill string, and the hands tong up the kelly to final tightness.

In this case, the kelly spinner replaces a spinning wrench, which is difficult to properly latch around the kelly and joint in the mousehole. Even better, a kelly spinner also replaces the spinning chain, the old-fashioned (and potentially dangerous) way of spinning up tubulars. To use a spinning chain, crew members wrapped the chain several times around the tool joint or kelly saver sub. The driller then used the automatic cathead on the drawworks to rapidly pull the chain from the pipe. As the chain came unwrapped from the joint, it spun the joint up. The chain could easily damage the tubular in the process of being wrapped and pulled off, as well as cause bodily harm if not properly used. Also, note that although it is acceptable to spin out a joint with the rotary, it is not good practice to spin up a joint with the rotary. Drillers should not spin up a joint with the rotary because when the shoulder of the rapidly spinning tool joint meets the shoulder of the stationary tool joint, the spinning joint suddenly stops. Although the spinning joint stops, the momentum of the spin goes down the drill string to the drill collars. The heavy drill collars do not turn as easily as the drill pipe; therefore, the turning force goes back to the drill pipe just above the collars. This turning force can be great enough to break out and unscrew a joint of pipe. Further, when the spinning joint stops, the momentum may be great enough to turn the pipe in the slips. As a result, the slip dies can score the drill pipe wall and create a place for failure.

Sometimes, it is more convenient to use the kelly spinner instead of the rotary for light-duty rotating jobs. For example, the driller may use the kelly spinner to rotate drill pipe slowly for fishing jobs or to orient directional drilling tools. Or, if the soil on which the rig's substructure rests is relatively soft, the driller can use the kelly spinner to drill a mousehole and a rathole. By making up a small bit on the bottom of the kelly, lowering the bit and kelly to the desired position under the substructure, and using the kelly spinner to rotate the bit, the driller can drill a mousehole and a rathole to the required depth, usually a matter of only a few feet, depending on the substructure's height and the length of the kelly and drill pipe joints.

Maintenance

Daily maintenance of a kelly spinner consists mostly of checking the lubrication system and greasing all lube points every trip. For a specific kelly spinner, consult the manufacturer's maintenance recommendations.

Problems that may arise when using the kelly spinner often have to do with a blockage of the air or hydraulic flow. Keep hoses straight, and check connections to be sure they are airtight. Replace valves and seals as necessary.

Spring Slips and Power Slips

Power slips (fig. 3) and *spiders* replace manual and spring slips to hold pipe in the hole while making and breaking connections. Modern power slips are a great improvement over earlier types. Power slips secure drill pipe and spiders hold casing; otherwise they work the same way: a gripping device—slips—that is usually actuated by the driller or another crew member, automatically suspends tubulars in the hole, thus relieving crew members from having to manually lift and place regular slips around the tubulars hanging in the rotary. Power slips are relatively expensive, but are becoming more common.

Lifting and setting slips for long periods can be hard work. A set of manual slips weighs anywhere from 200 pounds (91 kilograms) on up, so it takes two or three floorhands to properly position them by hand into the rotary opening. Further, if the pipe is not completely stopped before the crew sets the slips, they can pinch, or *bottleneck*, the pipe. Therefore, the driller on the brake and the floor personnel must coordinate carefully with each other. Near the end of a tour, after hours of picking up and setting down the slips and handling pipe, crew members can become tired.

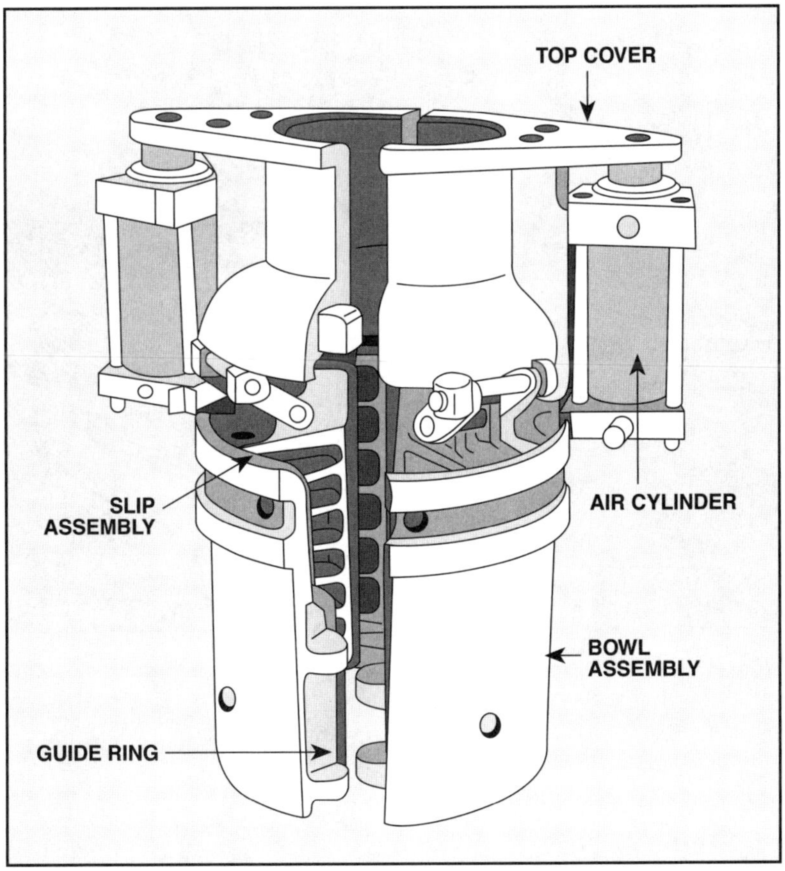

Figure 3. Power slips may be pneumatic, like this one, or hydraulic.

If they start dragging and letting the slips fall into the rotary opening under their own weight, instead of picking them up and lowering them properly, the falling slips can damage both the pipe and the slips.

Spring slips are an improvement over manual slips. Crew members set spring slips by standing on them (fig. 4). The slips unlatch automatically when the driller picks up the string.

Power slips can solve the problems of timing and mistakes caused by fatigue. With most types of power slip, the driller sets them and the crew never touches them. When tripping in or out, the driller stops the pipe, then steps on a foot control that sets the slips. After the connection has been made or broken, the driller uses the same foot control to move the slips up and out of the way. The foot valve may operate the slips either pneumatically or hydraulically through hoses that run under the rig floor.

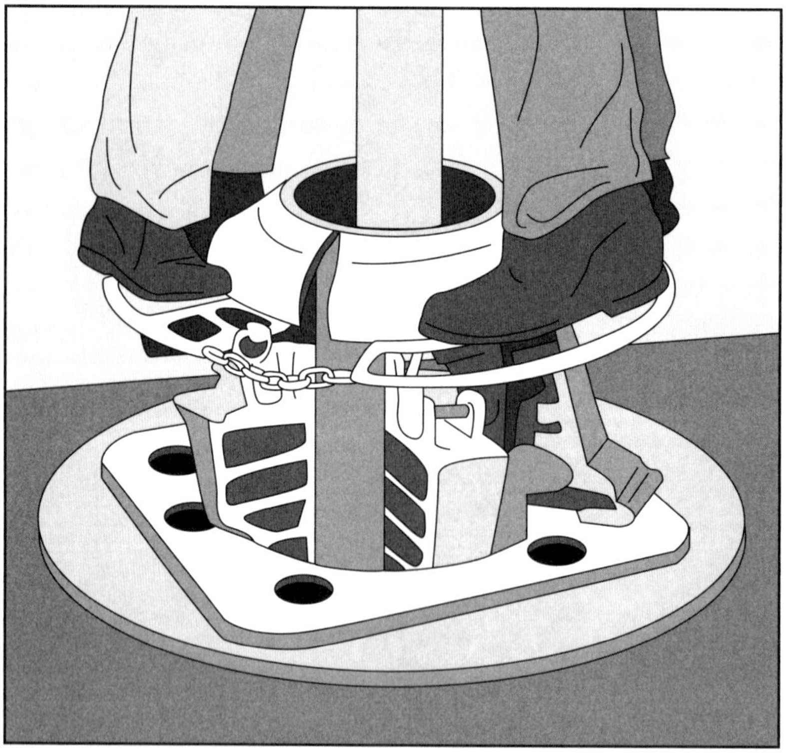

Figure 4. Two people stand on spring slips to set them.

Maintenance

As with most mechanical equipment, lubrication is the first rule of preventive maintenance. Follow the manufacturer's lubrication schedule, and grease everything that moves. Clean any abrasive material from the inside of the bowls and the outside of the slips (fig. 5a). Then lubricate these surfaces with grease to keep the slips from sticking in the bowls. Because abrasive drilling mud can cause horizontal lines of wear on these surfaces, rub an emery cloth vertically along these surfaces to smooth them out (fig. 5b).

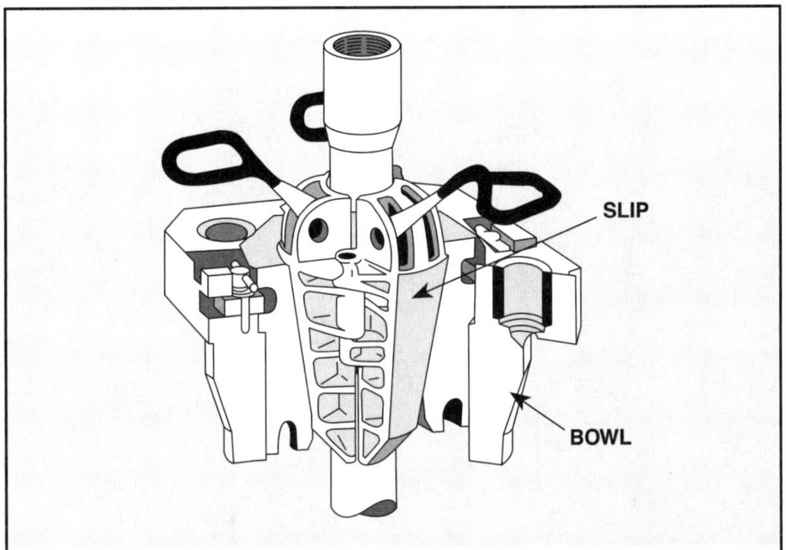

a. Clean and lubricate the inside of the bowls and the outside of the slips.

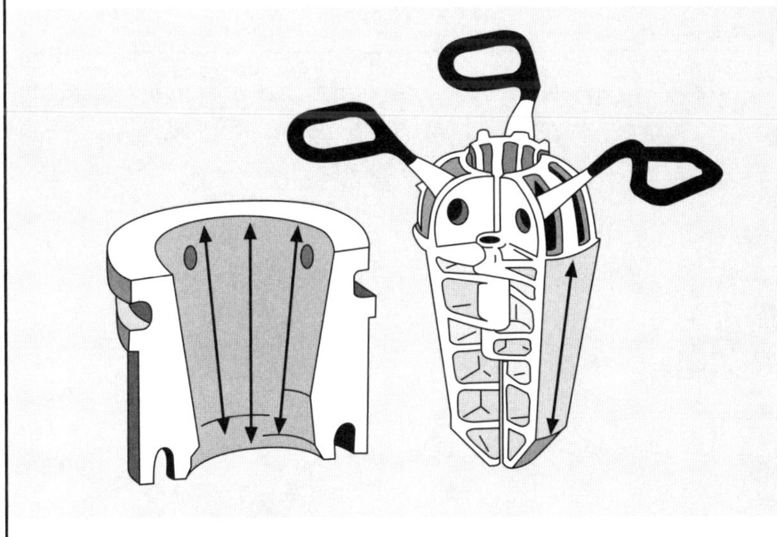

b. Dress the inside of the bowls and the outside of the slips by rubbing up and down with emery cloth.

Figure 5. Maintenance of slips

Inspect the slips for damage periodically. Put a straight-edge along the back and inside face of the slips to make sure they have not bent (fig. 6). Look for cracks, and destroy or get rid of cracked slips. Inspect the hinge pins regularly for wear, cracking, and straightness. A bent hinge pin means that the hinge pin hole has enlarged, and the slips must be replaced. Check the insert bowls and insert slots regularly for wear. Replace the slips when the clearance between the back of the insert bowls and the insert slots is smaller than the manufacturer recommends.

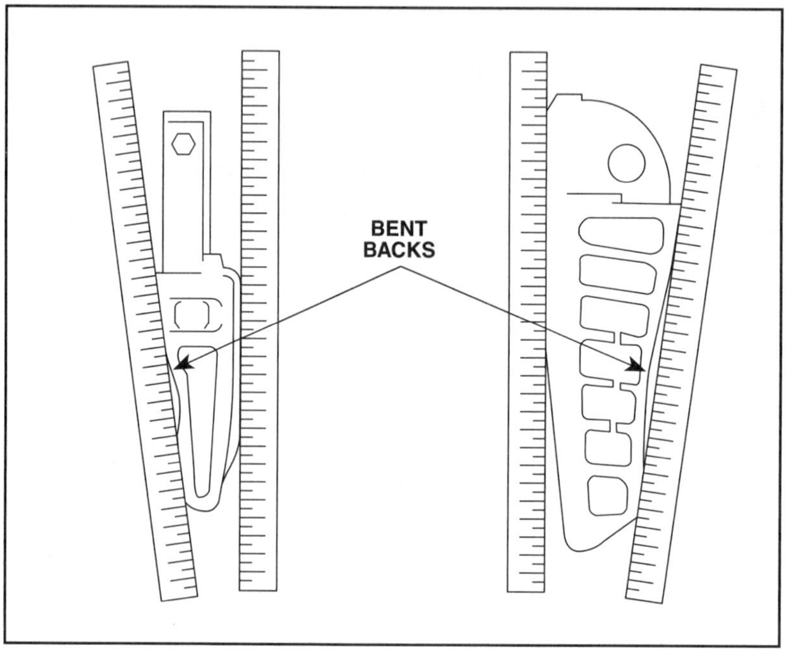

Figure 6. Check the slips for straightness with a straight-edge.

Automatic Pipe Handlers

With offshore rigs drilling holes in ever increasing water depths, the amount of time spent handling tubulars and tripping in and out can be much greater than on rigs drilling relatively shallow holes on land or in shallow water depths. The need to spend more time handling tubulars occurs not only during drilling, but also while running casing and riser pipe. More time spent on handling tubulars means less time making hole, so economics is one motive behind the development of automated pipe-handling systems.

Other issues include the potential danger and heavy labor involved with handling pipe. More rig accidents happen during pipe handling than any other operation. Therefore, systems that make it safer and easier to handle pipe is often an attractive option.

History

Efforts at automated pipe handling have been around since the early 1950s, when power elevators, power tongs, and semi-automated racking systems were introduced on land and offshore in the United States. The Soviet Union increased the use of mechanized pipe handling during the 1960s and 1970s.

In 1968, a U.S. manufacturer built a successful system that was the first to perform more than one critical operation at the same time: it moved the traveling block away from the well center so that racking arms could move a stand while the driller was raising or lowering the block. This system also allowed the racking arm operators to work inside climate-controlled cabins, away from the dangers of swinging pipe and block. Working in harsh environments became much less onerous. Research in the 1970s showed that a mechanized pipe-handling system saved 28 days per year of drilling and at least 1 hour per round trip.

By the 1980s, the trend toward complete automation was underway as new computer technologies became available. Top drives (power swivels) and *Iron Roughnecks*^{TM} were the first tools to be computerized so that the driller could control automatic sequences of action from the console. These tools, along with power slips, have given the driller much more control over the handling of the drill string. The next step was to integrate the top drive, power slips, and automatic Iron Roughneck with pipe racks and fingerboards. A growing number of rigs use mechanized and automated pipe-handling systems. At first, they appeared primarily on newly constructed offshore rigs. Now, however, companies are installing them on existing offshore rigs and some land rigs.

Pipe-Handling Systems

Automated pipe handling can be broken down into two aspects—horizontal pipe transfer systems and vertical rackers. As the names imply, one handles pipe while the pipe is horizontal and the other while the pipe is vertical. For both, the driller uses an integrated control system at the driller's console to move the pipe, leaving the pipe untouched by human hands.

Horizontal Pipe Transfer Systems

Horizontal transfer systems have several components that work together to move pipe from the pipe racks into position over the mousehole or well center (fig. 7). First, a pipe deck crane picks up the pipe from its horizontal position on the pipe rack and places it on a conveyor belt or rollers. The conveyor moves it through the V-door up to the rig floor. There, a *pickup-and-laydown system* tilts the pipe to vertical and moves it into place.

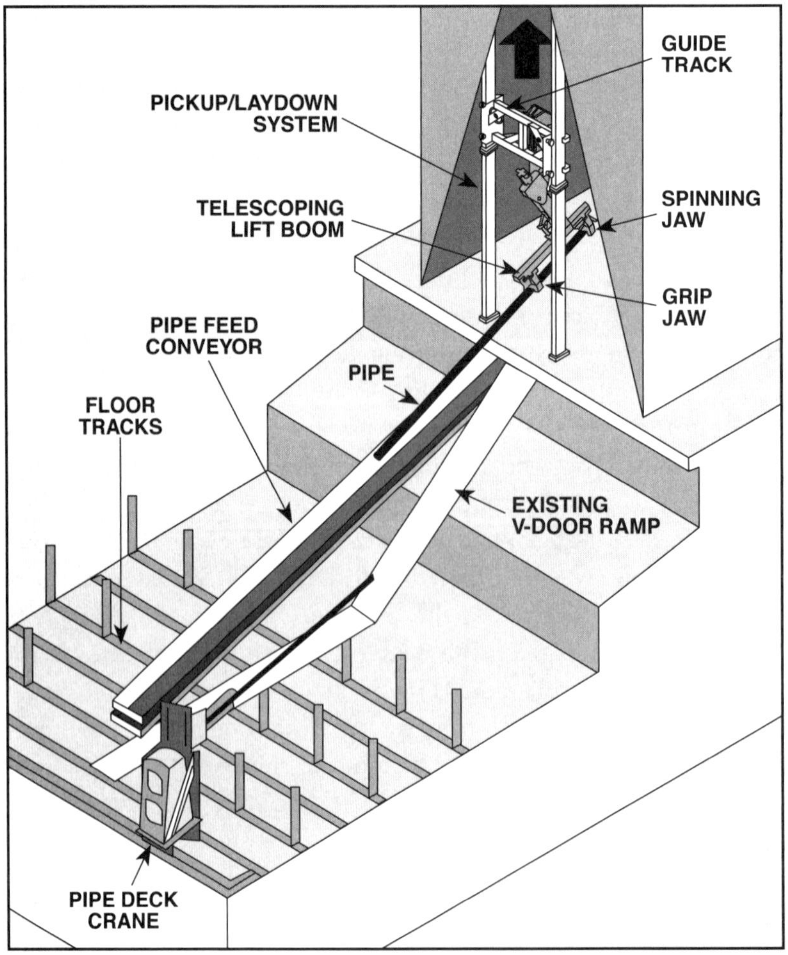

Figure 7. Pipe handling

Vertical Racking Systems

Vertical racking systems, or *column rackers*, are automated crane-and-rack systems that store pipe vertically on the rig floor and move it to the well center and back (fig. 8). The fingerboard holds pipe

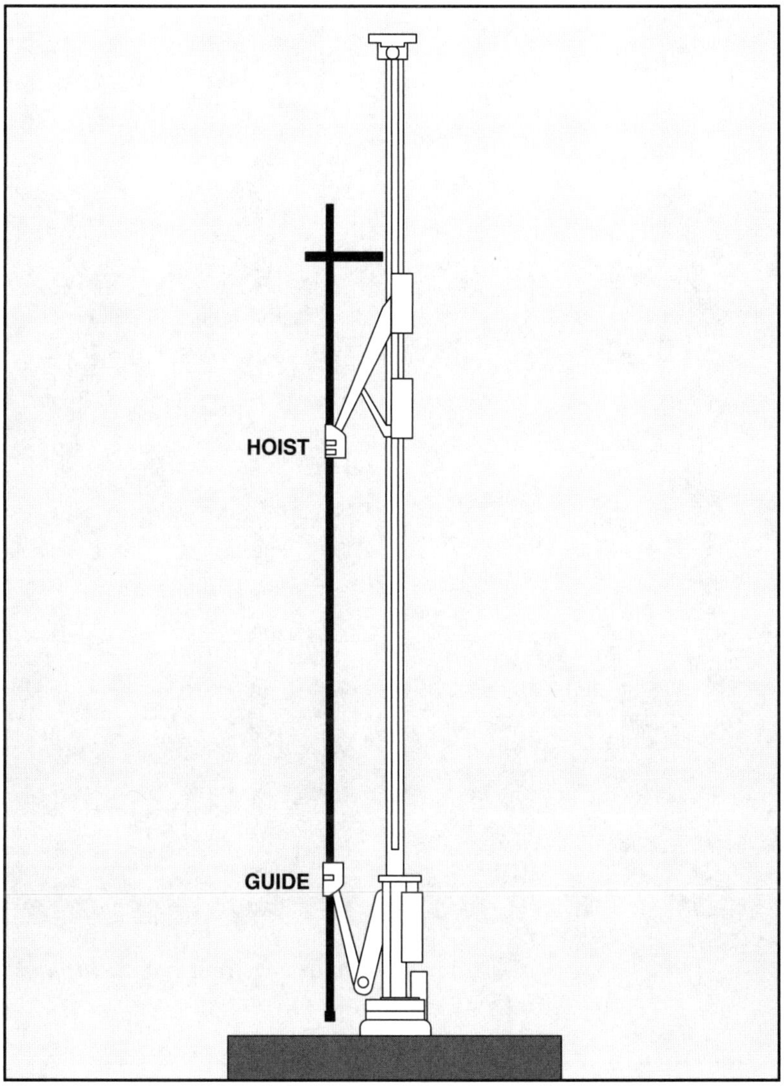

Figure 8. A column racker has arms that grip the pipe at both ends.

either in a parallel configuration (fig. 9a) or in a radial, or star, pattern (fig. 9b). The system consists of a crane attached to the floor with arms that latch onto the pipe at top and bottom to move it.

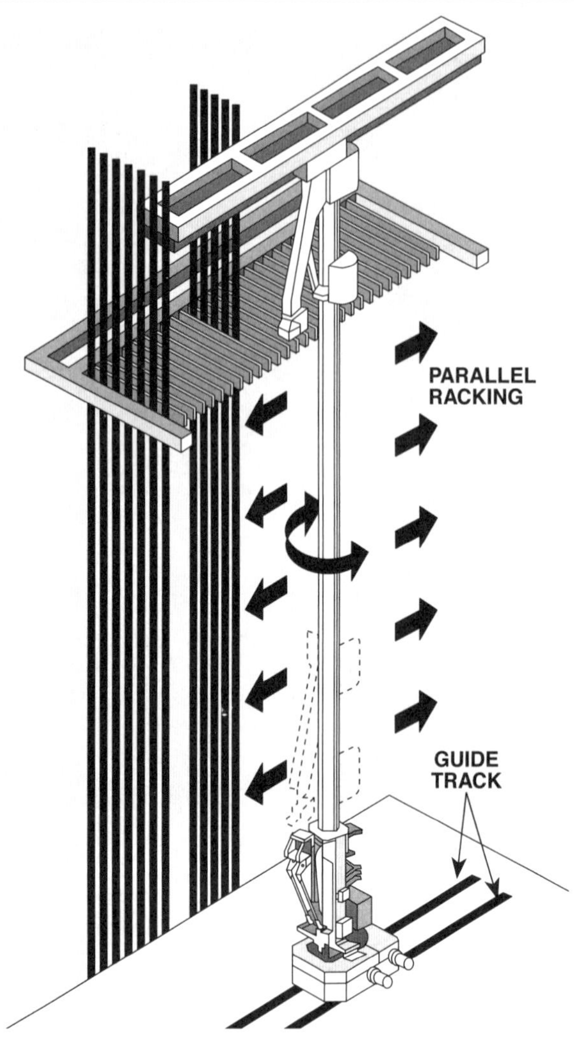

a. A parallel racker lines up the pipe in parallel rows.

b. A star racker has fingers that radiate around a center.

Figure 9. Fingerboards

Advantages

The most important advantage of automated pipe handling is increased safety. Automated pipe-handling systems replace floorhands using air hoists, cranes, and catlines. When crew members use a crane, for example, to move pipe from the pipe racks to the rig floor, the crane operator has limited control over the pipe's swinging or movement in the air. What is more, once the pipe enters the V-door, crew members must then further move the pipe into position on the floor, often using a hoist line to position the pipe. Again, the pipe can swing or move abruptly and endanger those working on the floor.-

---

To summarize—

- A kelly spinner is a motor mounted on the kelly that replaces the rotary and spinning chains to make up or break out pipe.
- Power slips and spiders allow the driller to set and release pipe in the hole with controls at the console, so that the crew never touches the slips or pipe.
- Automated pipe-handling systems allow the driller to move pipe from horizontal or vertical racks to the hole using controls at the console. This speeds up tripping, prevents accidents due to manual pipe handling, and frees the crew for other tasks.
- Pipe transfer systems consist of a pipe deck crane, a conveyor belt or rollers, and a pickup-and-laydown system to move pipe from a horizontal position on the pipe rack to a vertical position over the well center or mousehole.
- Vertical racking systems consist of a rack to hold pipe vertically and a crane to move it over the well center or mousehole.

Rig Instruments

A drilling rig instrument measures drilling parameters, equipment function, or formation characteristics; displays the measurements on a panel or a readout device; records the measurements; controls equipment within set limits; and stops operation if control fails.

Instruments on a rig include sensors, gauges, recorders, and various tools to control the machinery. Instruments can be bought and used independently of each other, but manufacturers also offer instrumentation systems where many instruments feed data to a central computer.

Sensors, Indicators, and Recorders

Sensors measure *drilling parameters*, which are factors that affect a drilling operation, such as the rate of penetration, pump rate, rotary revolutions per minute (rpm), weight on bit, and the like. The sensors send signals to an analog or digital readout, or gauge, which displays the information. An analog display is usually a needle on a dial (fig. 10). A digital display may be a liquid crystal display (LCD) or an electronic graphic representation on a cathode ray

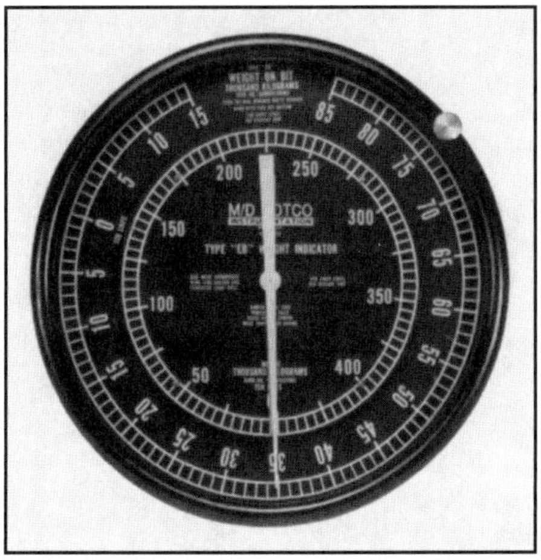

Figure 10. This weight indicator has an analog readout.

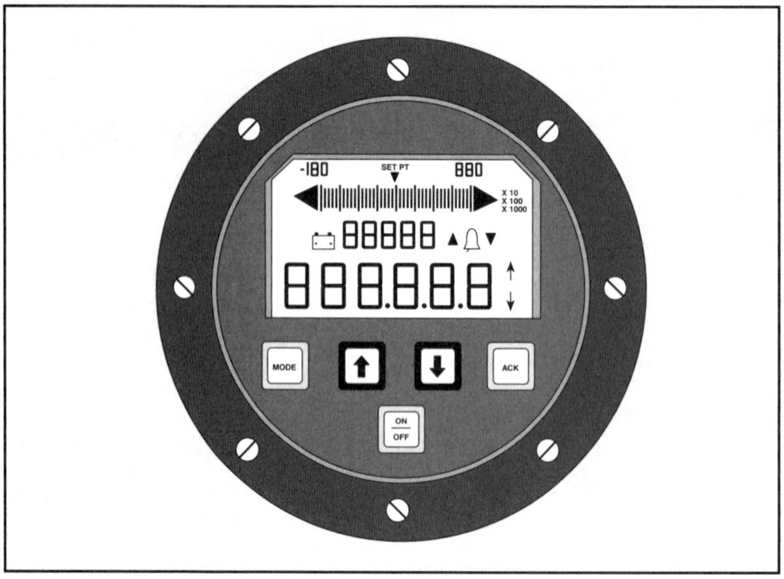

Figure 11. A digital readout uses LCD technology to show words and numbers.

tube (CRT), which is similar to a standard TV screen. In either case, the display shows numbers and words or graphs (fig. 11).

Sensors send the information they are monitoring through a pneumatic or hydraulic hose or an electric cable to the gauge. Analog gauges merely display the information. Digital gauges are usually much more sophisticated. Not only do they display information, they often have a microprocessor inside that interprets the information they receive. What is more, many digital systems include a keypad that allows the operator to set alarm limits, calibrate the sensor, and input other simple commands. Digital gauges usually do not require a great deal of maintenance because they have no moving parts to wear out and do not rely on high-pressure hoses as do pneumatic or hydraulic monitors.

Weight Indicators

Drillers need accurate measurements of *hook load* (the total weight that the blocks and drilling line support) to ensure that they are not exceeding the rig's lifting capacity. They also need an accurate measurement of *weight on bit* to control the rate of penetration and the direction of the hole. A weight indicator (fig. 12) provides these two measurements. A *weight indicator* is usually a relatively large analog gauge with two pointers. One pointer indicates hook load and the other weight on bit. Normally, the weight indicator is mounted at the driller's console. Weight indicators receive a hydraulic signal from a hydraulic load cell, which is a special disk-shaped piston-and-cylinder device that is clamped to the deadline. Hook load and weight on bit put tension on the deadline: generally speaking, the greater the tension, the heavier the weight and vice versa. The hydraulic load cell's piston moves with increasing or decreasing deadline tension and transmits the tension by means of hydraulic fluid contained in the load cell and in a heavy-duty hose running to the weight indicator. Variations in hydraulic pressure cause the pointers on the weight indicator to move.

Another type of weight indicator electronically senses deadline tension. A strain gauge, which is an electronic device called a transducer, converts force on the deadline to an electrical signal. (Transducers convert one kind of energy into another form; in this case, tension into electricity.) The electrical signal feeds a digital display, which indicates hook load and weight on bit.

Figure 12. Pointers on a weight indicator show weight on bit and hook load.

Wire Rope and Wireline Monitors

The rig may have a number of monitors for use with a wireline and wire rope, especially the drilling line. (Wireline is a slender, rodlike or threadlike piece of metal, usually small in diameter, that crews use to lower special tools, such a logging sondes, perforating guns, and the like, into the hole. Wire rope, on the other hand, is a cable composed of steel wires twisted around a central core of fiber or steel to create a rope of great strength and flexibility.)

Usually, drilling line wear is measured in ton-miles or, in the SI system, in megajoules. A drilling line does a ton-mile of work when it has moved a weight of one ton over a distance of one mile. Similarly, a drilling line does a megajoule of work when it moves a load representing 1,000 newtons over a distance of 1,000 metres. (In the SI system, the newton is a measure of force, not weight, or mass.) A ton-mile or megajoule indicator uses sensors to record how much weight the line has lifted or how much force has been put on the line and how many miles or metres it has traveled in either direction. Then it automatically calculates and displays line wear in terms of ton-miles or megajoules. The ton or newton sensor, which is a transducer, converts a pressure signal from the rig's weight indicator to an electrical signal, which an electronic computer chip automatically processes and translates into tons or newtons. A distance sensor measures the line's traveling distance by detecting the motion of the drawworks drum. It is mounted so that bolt heads on the drum, for instance, pass near the sensing head. Each ton-mile or megajoule indicator is calibrated to a specific drawworks, so they cannot be interchanged.

A wireline weight indicator shows operators how much a tool attached to the wireline weighs while downhole. Thus, a wireline weight indicator can show when the tool has come loose from the line because the weight on the line will drop to virtually zero. An aluminum tension load cell or a hydraulic load cell in the weight indicator's system sends information to an analog display through a hydraulic hose. As mentioned earlier, a load cell is a piston-and-cylinder device that senses changes in line tension, which indicates changes in weight on the line.

Rotary Torque Indicators

The amount of turning force, or torque, being put on the drill string either by the rotary table or a top drive is important to know. For example, if the rotary torque is greater than expected for a particular drilling job, the higher torque stresses the drill pipe. In extreme cases, the torque can be so high as to cause the drill string to break or twist off. Higher than normal torque can be caused by an undergauge hole, a change in the formation, or a problem with the bit, such as locked cones.

If the hole is smaller than the bit being used to drill it (if it is undergauge), then the bit's hard contact with the sides of the undergauge hole will increase the torque on the drill string. If the formation being drilled changes—for example, if it becomes too hard for the bit being used to efficiently drill it—torque will increase. Finally, a roller cone bit, which is characterized by usually having three cones that rotate on bearings as the bit rotates, can experience locked cones. That is, for various reasons, one or more of the bit's cones can stop rotating even though the bit itself continues to rotate. Locked cones increase torque on the drill string.

Instruments that measure rotary torque may be mechanical or electrical (fig. 13). A mechanical indicator is a load cell that is usually installed on mechanical rigs—rigs that use chains, belts, and sprockets to transfer engine power to the rig equipment. The load cell in this case measures force. The force is applied across a special substance, usually a crystal in the cell, that generates electric current proportional to the force. Crewmembers usually attach the mechanical load cell under the rotary drive chain,

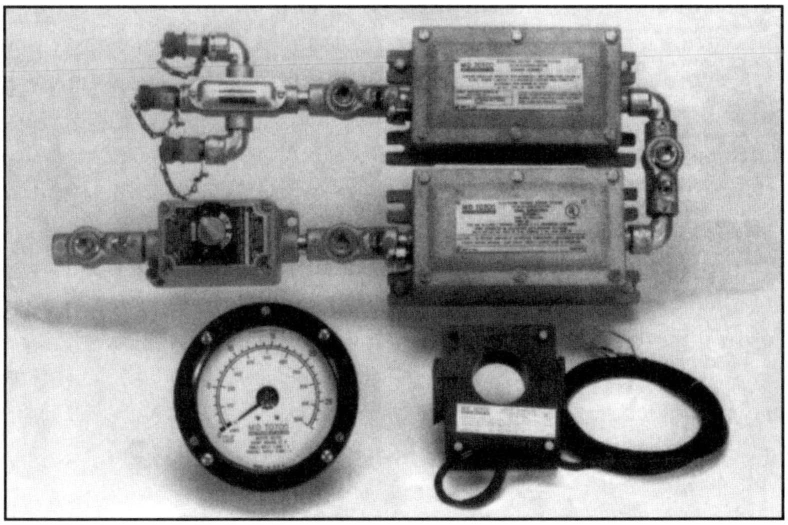

Figure 13. This electric rotary torque indicator uses a transducer to monitor torque on an electric rig.

where it senses changes in the chain's tightness. In general, as torque increases, the chain tightens; as torque decreases, the chain loosens. The load cell sends its signal to a gauge on the driller's console.

An electric torque indicator is usually employed on electric drive rigs and on top drives. Crew members clamp a transducer around the power cable to either the rotary drive motor or the top drive's motor. The transducer measures electrical current put out by the motor as it turns the rotary or the drill stem. Since the amount of current flowing from the motor is an indication of the amount of torque, the transducer translates current flow to torque and transmits the information to a readout or a gauge on the driller's console.

Besides rotary torque, another kind of torque is that which is required to make up tool joints, drill collar connections, and other tubular connections. Thus, a torque indicator is available that shows the amount of makeup torque being applied to tubular connections by the tongs or other devices being used to tighten the connection. With one kind of torque indicator, the driller can set the torque indicator's readout gauge to the desired torque, and then watch the gauge to make sure that this torque is not exceeded as crew members tighten the joint with the tongs.

RPM and SPM Indicators

It is necessary for the driller to know the speed at which the rotary table is turning, as well as the speed at which the mud pumps are pumping. Different types and sizes of bit require different turning speeds to drill the most efficiently. Pump speed is also critical to efficient drilling because it affects how well the mud cleans the hole of cuttings made by the bit. Also, during well-control operations, well-control personnel must know the pump's speed to properly kill the well and regain control. Rotary speed indicators usually measure the speed of the rotary table in revolutions per minute (rpm). Pump speed indicators usually measure pump speed in strokes per minute (spm).

An rpm indicator may use a magnetically activated probe next to the rotary table to measure speed. All iron-based metals, such as the steel from which rotary tables are made, can be magnetized.

While the magnetic field may be small, it is sufficient to enable an electromagnetic probe to sense variation in the magnetic field that occurs as the table rotates. The probe translates the variation into an electric indication of rpm. This indication then goes to an rpm gauge, which gives a readout. Another type of rpm indicator includes a small generator that is attached to the rotary table's drive shaft or to any other shaft in the rotary machine that turns in direct proportion to the rotary table. The generator's turning speed varies with shaft speed, which varies with rotary table speed. When the generator turns, it puts out electric current that carries the rpm reading to a gauge near the driller's position.

A common type of spm indicator consists of a hinged, mechanical probe mounted on the body of the fluid end of the pump. The probes are positioned so that they are close to the pump pistons. As the pistons move back and forth, they strike the probe, which moves back and forth on the hinge mechanism. With each back-and-forth movement, the probe sends an electric signal to an spm gauge mounted at the driller's position.

Another type of spm indicator is mounted on the pump pinion shaft, piston rod oiler, or the V-belt driving the oiler. A small generator turns in proportion to the speed of the pinion shaft, oiler, or V-belt and sends a signal to the spm gauge on the driller's console. Because the shaft, oiler, or V-belt turn in direct proportion to the pump's stroke, the generator measures the pump's speed in spm.

Maintenance and Calibration

In general, indicators and sensors do not require much maintenance. It is important, however, to periodically check all fittings and hoses for tightness and for damage and to promptly replace worn or damaged parts.

As for calibration, most are calibrated by the manufacturer and therefore do not require any adjustments in the field. Weight indicators are, however, an exception. The driller should calibrate the rig's weight indicator after each move or whenever there is reason to suspect that it is not reading accurately. An adjustment knob is provided on the instrument, which the driller can use to calibrate it while following the manufacturer's printed instructions.

Recorders Recorders may be electronic or mechanical. A mechanical recorder, often called a Geolograph, which is the brand name of one kind of mechanical drilling recorder, has from two to eight pens that mark a moving roll of paper to produce a strip chart (fig. 14). The strip chart is a special roll of ruled and labeled paper, which, when read after the pens make their marks, shows a 12- or 24-hour record of the function, or parameter, the recorder is measuring. Depending on the number of pens, a recorder creates a chart, or log, of one or more drilling parameters—for example, drill string weight, weight on bit, penetration rate, rotary speed, rotary torque, pumping rate, fluid pressure, mud pit level, fluid flow rate, or straightness of the hole. Electronic recorders show the same type of information on a computer screen (fig. 15). Hard copies of the charts can be printed out as needed. This type of recorder is often part of an integrated drilling system, which is discussed later.

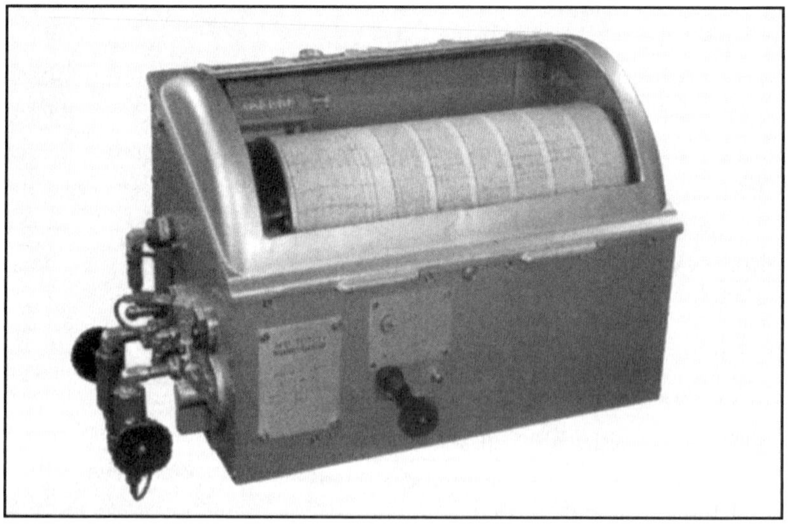

Figure 14. A Geolograph recorder produces a 12-hour or 24-hour strip chart using 2, 4, 6, or 8 pens.

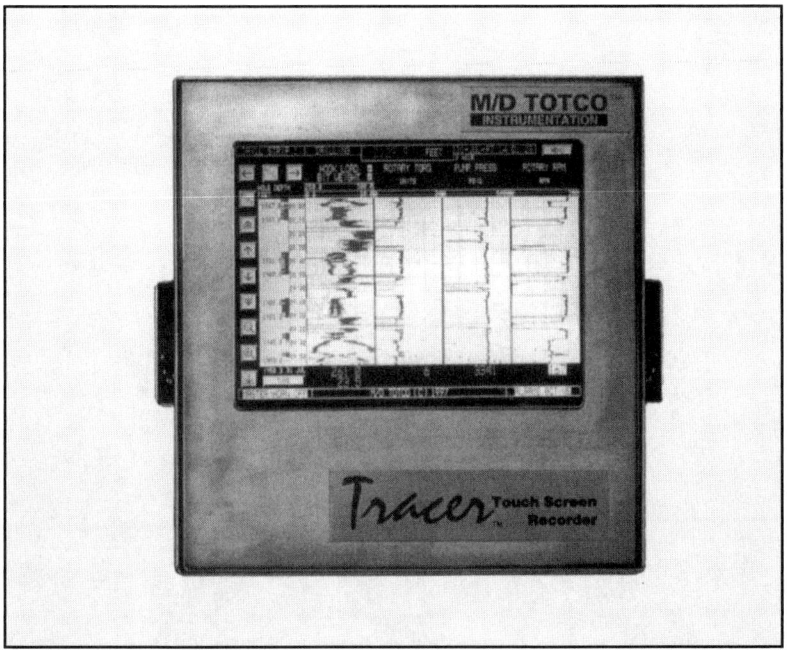

Figure 15. The driller can touch the screen of this electronic recorder to change which parameters are visible, scroll through strip charts, and enter notes.

Drilling Tools

▼
▼
▼

Adjustable Choke

A drilling choke allows personnel involved in controlling a kick to maintain a predetermined amount of back-pressure on a well while circulating the kick out of the well. A kick can occur when the pressure in the hole opposite a porous and permeable formation is less than the pressure of the fluids in the formation. When pressure in the hole is less than formation pressure, the hole is said to be underbalanced. When the hole is underbalanced, formation fluids can enter the hole. Pressure in the hole can be less than formation pressure when the weight, or density, of the drilling mud is not great enough to develop enough pressure to balance formation pressure. An alert drilling crew promptly notices that a kick has occurred and takes steps to control the well—that is, to prevent further entry of formation fluids and to increase pressure in the wellbore to balance formation pressure.

Actions crew members take on noting a kick include stopping mud from circulating by stopping the mud pumps and shutting in (closing) a blowout preventer, which closes in the well and prevents the further entry of formation fluids. But they cannot open the blowout preventer until the mud weight has been increased and circulated throughout the well; otherwise, formation fluids could re-enter the well.

Figure 16. Choke and choke manifold

To circulate a shut-in well, an adjustable choke in a special manifold (a series of piping and valves) is usually employed. A choke line is installed in the blowout preventer stack so that, when a valve is opened in the choke line, well fluids can be circulated through the choke manifold and choke (fig. 16). To circulate the well and hold enough back pressure on the well to prevent further entry of formation fluids, the operator starts the mud pump while simultaneously opening the choke. Well fluids (mud and kick fluids) flow through the choke. The mud goes back into the circulating system after kick fluids are removed from it. The size of the choke opening determines the fluid pressure—if the pressure starts to drop too much, the operator decreases the choke's opening, which increases the pressure. Conversely, increasing the choke's opening decreases the pressure. Instruments on a choke control panel show pressure, the pump's spm, and the choke position.

Automatic Drillers

An automatic driller helps the human driller keep the proper weight on bit (WOB). One of the driller's jobs is to maintain the WOB needed to efficiently drill a particular formation. Without an automatic driller, the human driller must constantly adjust WOB by manipulating the drawworks brake handle. The driller releases the brake a small amount to increase WOB as the bit drills down. Then, as the bit bites into the formation with the increased weight, the driller applies the brake to keep too much weight from being applied to the bit. Thus, when the bit is on bottom and drilling, a human driller must maintain constant attention to the drawworks brake. As the bit drills down, the weight decreases because the brake is set and does not allow the drawworks to let out any drilling line. An automatic driller uses a the weight indicator to sense the amount of WOB and automatically adjusts the brake to maintain WOB within fairly close limits. On the instrument in figure 17, the driller turns the dial in the center to adjust for more or less WOB, and the dial on the left to control how fast the brake is applied or released. A controller operates the brake pneumatically or hydraulically.

Figure 17. An automatic driller uses pneumatic power to help the driller keep the proper weight on bit.

MWD and LWD Tools

(LWD) are systems that give the driller information from the bottom of the hole. They save time and expense because the driller does not have to stop drilling to find out what is going on downhole.

How They Work

Measurement while drilling (MWD) and logging while drilling (LWD) tools use the drilling mud as a medium to carry information. The system works something like a stereo system. In a stereo system, electrical impulses cause speakers to vibrate, which produces sound waves in the air. These sound waves carry the information from the electrical impulses to our ears, and we hear a guitar or a bagpipe or an orchestra of different sounds.

In MWD and LWD, a battery-powered pulse generator inside the drill collars near the bit emits pulses that create waves in the drilling mud, similar to sound waves in the air. The waves move up the fluid column inside the drill stem. The waves carry information about the direction the bit is drilling, hole depth and angle, mud temperature and pressure, downhole WOB, downhole torque, and, in the case of LWD instruments, electrical, acoustic, and nuclear properties of the formations. Sensors on the surface pick up the waves and send the information to a computer. The computer analyzes the information and presents it visually.

Uses

MWD is most often used in directional and horizontal drilling. The MWD's readout shows the direction and angle of the bit as it drills and allows the driller to steer the bit to keep it on course.

Because LWD instruments make measurements at the bit, rather than 60 to 100 feet (18 to 30 metres) uphole, the driller has information about the geology of the formation at the time it is being drilled through. This real-time information allows the driller to guide the bit geologically rather than according to a predefined plan. In addition, the computer software can recommend the best path for drilling and the best mud hydraulics and jar placement.

Integrated Drilling Systems

When a rig is equipped with conventional drilling instruments, the readout for each instrument is usually mounted on the driller's console (fig. 18). With conventional instruments, the driller, in order to maintain optimum drilling conditions, must continuously look at all the information, from mud flow rate, to WOB, to ROP. The driller must then determine how all the information interrelates and affects drilling efficiency. Mistakes can be dangerous, costly, or both.

In an integrated drilling system, a computer helps analyze the information from the instruments and presents it in a way that allows the driller to make decisions more quickly and accurately.

Figure 18. The driller's console has space for many analog and digital readouts.

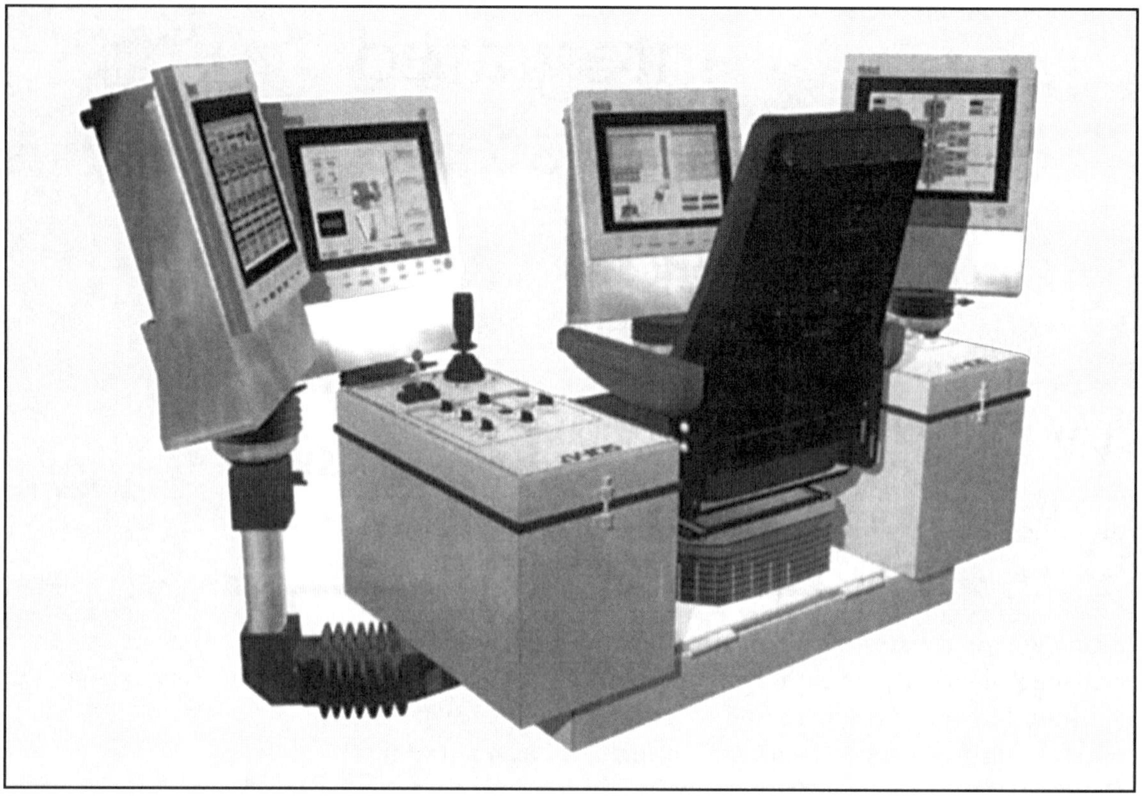

Figure 19. The driller sits in an ergonomically designed chair inside a climate-controlled cabin, and controls drilling through a joystick, switches, and touch screens.

The screen monitors at the driller's console show everything the readouts do and more (fig. 19). Also, the toolpusher and company employee may each have screen monitors as well, so they see the same information at the same time and can coordinate with the driller. In the figure, the joystick at the left of the chair controls the brake. A wheel on the right controls the speed of the block. The driller can enter information to the computer by touching the screen; no keypad is necessary. These new workstations are ergonomic—that is, the placement and design of controls, screens, and the chair should prevent or reduce fatigue and injuries from repetitive movements. They are housed inside a heated and cooled cabin with plenty of windows for good visibility of the rig floor and derrick.

Because each well is different and because each contractor and operating company has individual requirements, integrated drilling systems vary from rig to rig. Usually, however, every system will include the special sensors, recorders, and software. Also, regardless of exactly which parameters the system controls, the computer is at its heart.

Computer

The computer in an integrated drilling system is a personal computer (PC), the same type of computer you may have in your home. It receives data from all the instruments in an electronic form (fig. 20). It then analyzes the data, makes sure they fall within safe limits, and sounds alarms if a problem is imminent. The computer can also control automated equipment. For example, it may steer a downhole motor. What is more, computers can be configured to diagnose problems and present possible solutions. Another advantage of a computer is that it can save all the information it receives and display it on demand for future use.

Instrument readouts that are displayed as charts on the computer's screen can be printed out to give strip charts that resemble those from a Geolograph. Further, computer software is available that allows the driller to add additional information to the charts as needed, such as the position of the bit and traveling block, connection times, and amount of pipe tripped in or out.

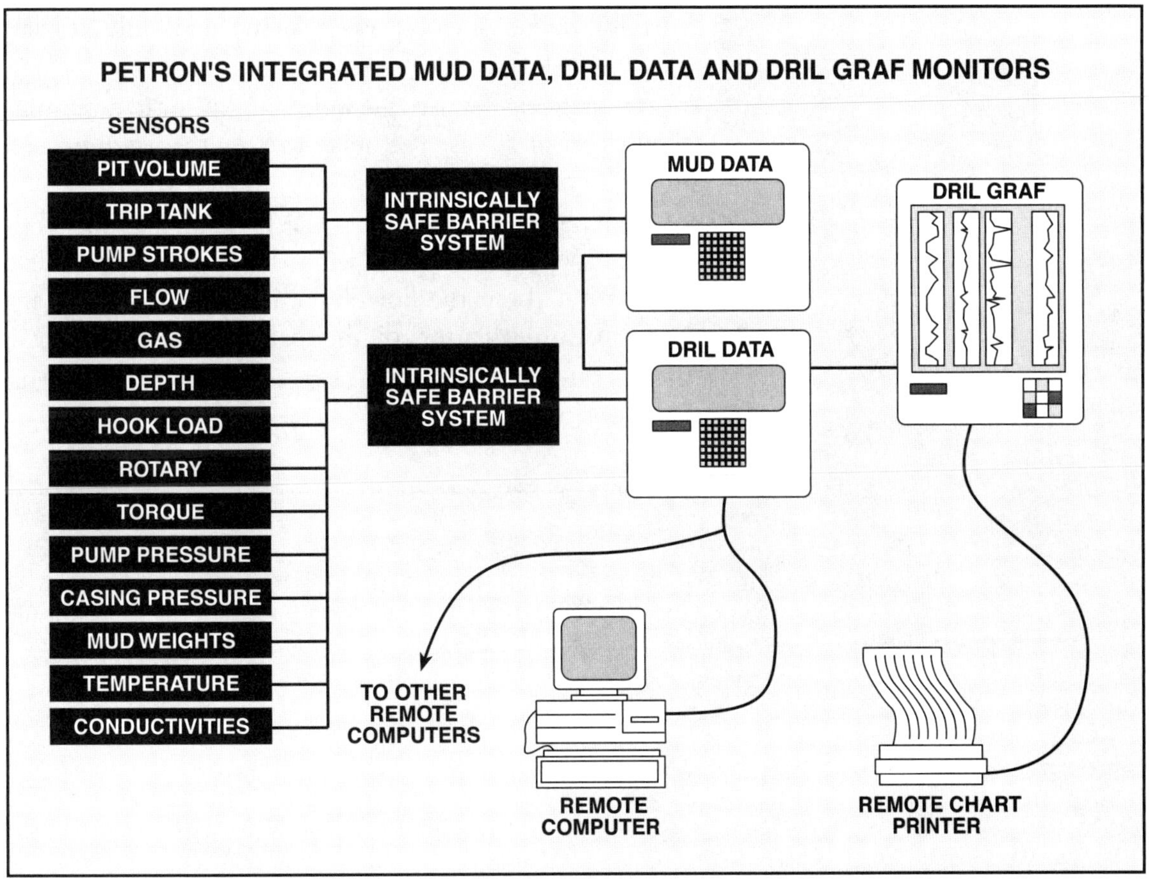

Figure 20. An integrated drilling system takes data from many sensors and routes it to readouts and computers.

Software is also available that lets the driller pose "what if" questions. For example, the driller can ask, "What will happen to the rate of penetration if I change the rotary speed?" The computer then analyzes data from the drilling sensors and displays an answer based on the data.

Through the computer, the driller, toolpusher, and company representative on the drilling site can connect with experts in offices far from the rig via the Internet or an Intranet (a private, company network). Sometimes a problem that stumps onsite experts can be solved by the fresh view of offsite experts.

To summarize—

- Sensors measure drilling parameters and send the information to analog or digital gauges. They may also send the information to recorders.
- A drilling choke controls pressure in the hole while circulating out a kick.
- An automatic driller uses the weight indicator to determine when to set or release the brake to keep the weight on bit within desired limits.
- Measurement while drilling (MWD) and logging while drilling (LWD) tools use pulses in the drilling mud to send information about the downhole drilling parameters and the formation to sensors on the surface.
- An integrated drilling system uses a computer to link instruments and analyze data.

▼
▼
▼

Utilities

▼
▼
▼

A drilling rig, like any other isolated plant or factory, demands the convenience of various utilities—fuel for the engines, water for auxiliary equipment and for human use, and compressed air and hydraulic systems to power auxiliary equipment. Electricity also powers much auxiliary equipment. Since most modern rigs are diesel-electric, the generators not only power the motors to drive the equipment, but also provide electricity to light and perhaps heat or cool the rig. Mechanical rigs require auxiliary electric generators, often called light plants, to provide power for auxiliary equipment.

Fuel Systems

Fuel systems may provide natural gas, liquefied petroleum gas (LPG), gasoline, diesel oil, crude oil, or any combination of these fuels. Today, except in rare instances, most engines that power the rig run on diesel fuel. Diesel is easier to transport and store than natural gas or LPG and it is not as volatile as gasoline. In a few instances, however, where a rig is operating near an easily tapped and abundant supply of natural gas or LPG, these fuels may be used. Similarly, in a few instances, where a small engine is required to operate a piece of auxiliary equipment, the only type available may be fueled by gasoline.

Natural Gas and LPG

If natural gas is the basic fuel for the rig's main engines, it can probably serve all rig fuel needs. Gas requires pressure regulation through one or more stages. Equipment for this purpose must be the correct size and must be protected against the possible formation of hydrates that could plug fuel lines.

An important consideration when using natural gas or LPG is that the rig must have a tank large enough to supply the engines with plenty of low-pressure gas when surges of power are required. Particularly with LPG, the supply must be continuous.

Gasoline and Diesel Oil

Gasoline is rarely used except for fueling small engines. Diesel oil is the most common fuel. Diesel oil is usually stored in large supply tanks at the drilling site. In addition, the rig usually has a small supply tank (a *day tank*) near the engines (fig. 21). A day tank is a relatively small diesel fuel tank located near the engines and situated such that it is easily filled from the main supply tanks. The day tank is usually required to facilitate rigging up. Because it is relatively small, it can easily be set up on the site, filled with diesel, and used to power the engines during rig up, before the main tanks are placed on the site. After rig up and the main supply tanks are set, they supply fuel to the day tank, which, in turn, supplies fuel to the engines. An essential item in a diesel oil system is a dependable supply pump, often an electrically driven gear-type pump. Level control switches on the day tank operate the pump automatically. (For more information on diesel engines, fuel systems, and related equipment, see Unit I, Lesson 8, *Diesel Engines and Electric Power*, which is available from Petroleum Extension Service.)

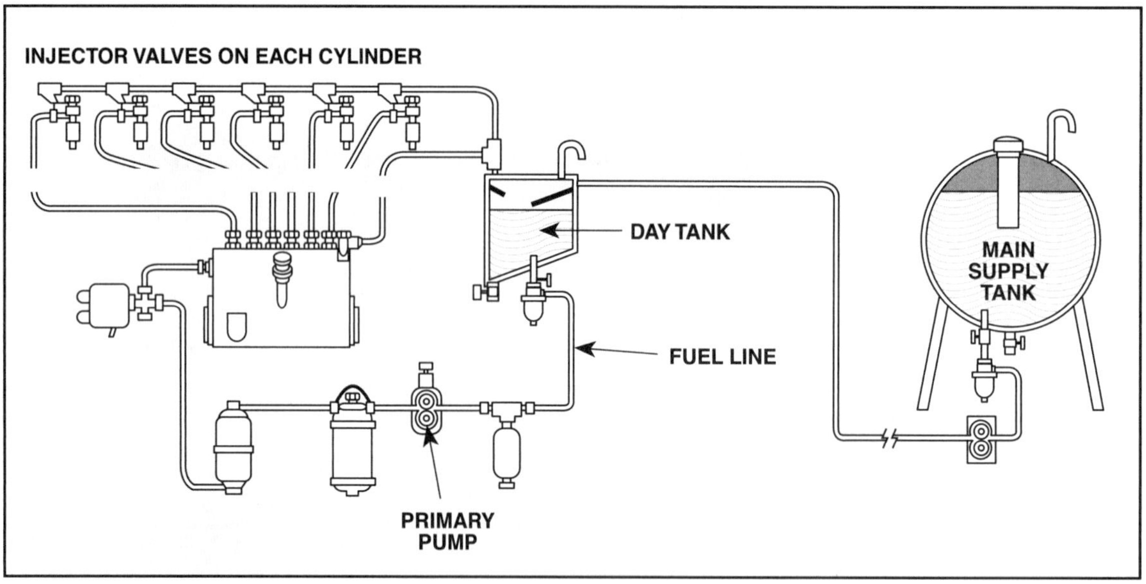

Figure 21. A day tank and pump supply diesel fuel to the rig.

Auxiliary Power Systems

Compressed Air

Compressed air powers a considerable amount of equipment on the rig. For example, some clutches and various other rig controls require compressed air to operate. Other pneumatic equipment includes the BOP accumulator pumps, kelly spinners, power tongs and power slips, and catheads. Air hoists and pneumatic tools are also operated by air. Further, pneumatic handling systems transfer dry bulk materials for drilling mud and cement. And motors powered by compressed air start engines and perform other tasks.

Many mechanical rigs have one or more two-stage compressors mounted on top of the compound housing, driven by a V-belt from the transmission. Electric rigs have an independently driven compressor (fig. 22). Special reinforced hoses (pneumatic lines) deliver the compressed air to the equipment that needs it.

Figure 22. An air compressor supplies pneumatic power to tools and auxiliary equipment.

Hydraulic Power

Water or, more often, another low-viscosity liquid such as light oil, can also power equipment, because liquids do not compress when energy is added. Instead, they transmit pressure. Rigs may have a permanent or portable source of hydraulic power fluid, at a pressure of about 2,000 psi (13.8 MPa), that flows through hoses. Power tongs and power swivels are the main users of hydraulic power, but hardly the only ones. Blowout preventer units have a separate hydraulic system, but in emergencies, the rig's system can serve this purpose.

Water Systems

On land, the rig's water system normally provides water for all uses except for human consumption. Often, rig water for mixing mud and the like comes from a well drilled specifically for the purpose or perhaps from a nearby stream, pond, or lake. As for potable (drinking) water on land sites, it is easy to truck in and store on the site. In isolated areas and offshore, however, a water system may not only include water for rig use, but also often includes equipment to produce drinkable water from some sort of raw stock such as seawater or brackish (salty) water. Filter purification units or evaporation units called *watermakers* are common in offshore operations, where the available water is salty seawater, and in desert areas where the only water available is from deep wells and is often brackish.

Storing and Transferring Water

Freshwater storage capacity at the drilling site may vary from a few hundred barrels (a few dozen cubic metres) to several thousand barrels (a few hundred cubic metres), depending on the water source and how quickly it will be consumed. Freshwater for mixing with drilling mud usually flows by gravity directly from bulk storage. The quality of the water that goes to radiators for engine cooling must be controlled; this water may require filtration and chemical treatment. When heat exchangers cool the engines, the composition of the supply water is less critical. A day tank set up beside the rig supplies water for cooling the brakes and for radiators. A pump fills the day tank from bulk storage.

Watermakers

Because offshore drilling and production rigs and ships are far from the usual sources of water, they often have a watermaker. A watermaker is a device that turns seawater into fresh water. Two types used in the oil industry are evaporators and reverse osmosis watermakers.

Evaporators

Evaporators use an old and simple technology to desalinate water. When salt water boils, the water evaporates and rises as steam, or water vapor, while the salt remains behind. Cooling the water vapor causes it to condense into pure water.

Figure 23 is a simplified diagram that shows how an evaporator works. Seawater is pumped into the evaporation chamber. Here it is heated to the boiling point, and the water vapor rises and enters a demister. The demister removes any saltwater droplets from the steam, and then the vapor flows through a pipe into a condensing chamber. The vapor hits pipes carrying cold seawater, and it condenses into pure water, or distillate. The water that remains behind in the evaporation chamber has more salt than seawater. This brine is continuously replaced with seawater. Many kinds of evaporator are available and, while they vary somewhat, they all have the same basic components.

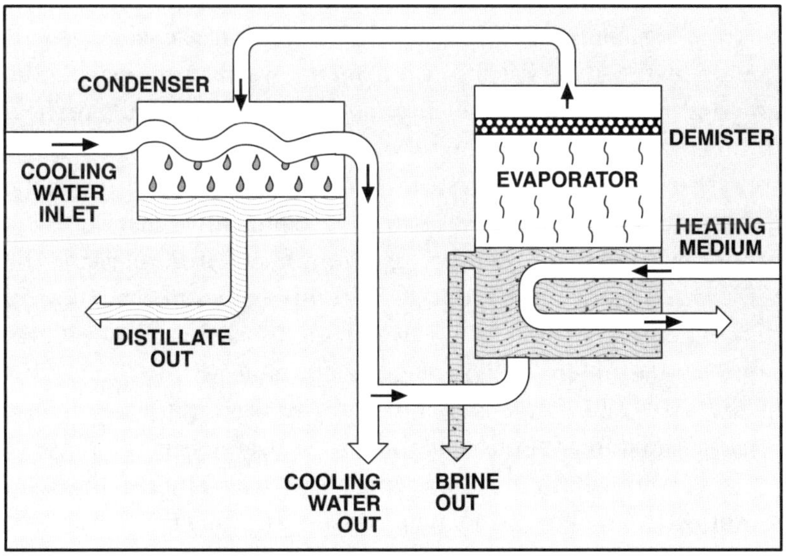

Figure 23. This simplified diagram is typical of how evaporators work.

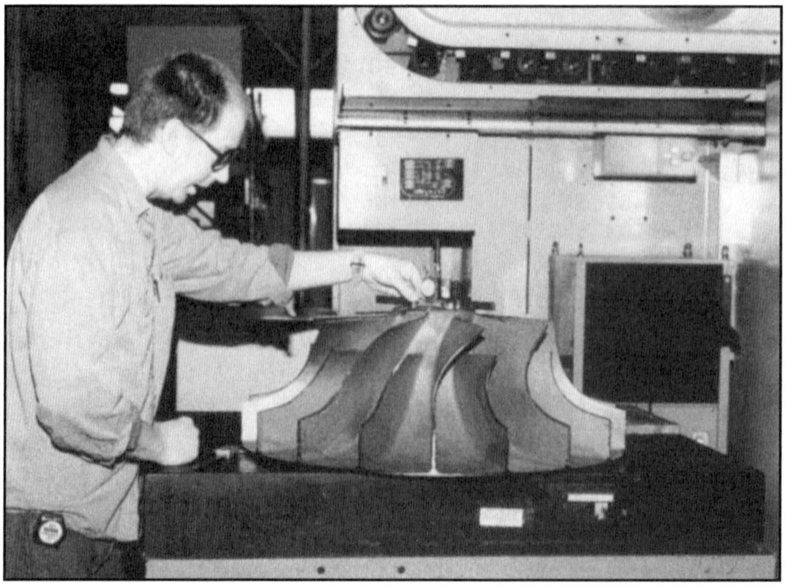

Figure 24. A vapor compression evaporator compresses the water vapor before condensing it.

Vapor Compression Evaporators. A vapor compression evaporator (fig. 24) uses electric motors to power pumps, a compressor, and a lubricating oil pump. Depending on the size of the unit, it can produce 4,800 to 16,800 gallons (about 18 to 64 cubic metres) of purified water per day.

A vapor compression evaporator has a heat exchanger—a titanium plate—that absorbs heat from the waste water leaving the unit and transmits it to the entering seawater. Transmitting heat in this manner preheats the seawater. The preheated seawater is sprayed over tubes carrying steam from a boiler. It rises through the demister. It then flows into a centrifugal compressor that raises the vapor's pressure. Like uncapping a hot car radiator, releasing the pressurized vapor blasts it out of the compressor and onto tubes in the condenser carrying feed seawater, on its way into the evaporation chamber. Here the vapor gives up its heat to the sea-water in the tubes and condenses. This type of evaporator is efficient because the water vapor heats the feed seawater as it condenses.

A centrifugal compressor works like a jet engine. It uses a motor or an engine to rotate an impeller (fig. 25) that turns on a drive shaft inside a housing, or casing. The rotating impeller forces the water vapor that enters the compressor out and away from the blades. The energy of the rotating impeller transfers to the water vapor, which increases its pressure and temperature.

Figure 25. The compressor's impeller is made of titanium or a corrosion-resistant nickel-chromium alloy.

Vapor compression evaporators must be started manually, but they then work automatically until shut down manually or until automatically shut down by a safety shutdown system that senses trouble. Valves control the flow rates and the level of distillate in the boiler. Strainers in the feedwater line filter sand and contaminants. Flow indicators, pressure gauges, and thermometers indicate conditions in the unit. A programmable logic controller on a computer chip monitors pressures, water levels, and lube oil temperature. The logic controller sets off an alarm and automatically shuts down the unit if the pressure, water level, or oil temperature fall outside safe limits.

Vapor compression evaporators require a good deal of maintenance. The main maintenance headache is scale. If you have ever boiled tap water in a pan, you have seen *scale*, a hard white residue on the side of the pan. It is made up of dissolved minerals in the water that do not evaporate. Injecting a chemical solution (a scale inhibitor) into vapor compression units can limit scale formation. Flushing the unit with cold seawater whenever it is shut down also helps reduce scale. Eventually, however, it must be cleaned with a strong acid. Since most mineral scales are composed of calcium and magnesium compounds, acids readily dissolve them. A container on the evaporator feeds the acid through an eductor with valves and piping to transfer it into the unit.

What is more, regularly check the motors, pumps, compressor, gearbox, and lubrication system. Lubricate all moving parts of the pumps and compressor in accordance with the manufacturer's recommended schedule. Also, check the belt that connects the motor to the compressor's driveshaft for proper alignment.

Waste Heat Evaporators. A waste heat evaporator (fig. 26) uses heat from the rig's diesel engines to boil seawater. The engine's cooling water pump circulates heated cooling water through the heating surface of the evaporator and then back to the engine's cooling system. This heat is not enough to boil water at normal (atmospheric) pressure, but, when put under pressure lower than atmospheric, water boils at a lower temperature than 212°F (100°C). So, a waste heat evaporator operates under pressure below that of the atmosphere; in other words, it creates a vacuum. Because it operates under a vacuum, the system only has to heat the seawater to about 115°–120°F (45°–50°C) to make the water boil. This type of evaporator, depending on its size, can produce from 100 to 800 gallons (about 380 to 3,000 litres) of fresh water per hour. Note that because low-temperature operation reduces the amount of scale that forms, acid cleaning is necessary less often than with a vapor compression evaporator.

Figure 26. A waste heat evaporator uses low pressure and waste heat from the rig's engines to boil seawater.

Flash Watermaker. A flash evaporator uses heated plates to rapidly vaporize seawater. It is smaller than a vapor compression evaporator and produces less water per day—only about 630 to 4,200 gallons (about 2,400 to 16,000 litres) per day. Some units use waste heat from the diesel engines as the heat source. An advantage of the plate flash watermaker is that the flash process leaves almost no scale, eliminating the need for chemical or acid treatments. Its pumps require the usual maintenance and lubrication. Like all equipment meant for offshore use, its parts are made of corrosion-resistant steel, copper-nickel alloy, or bronze.

Reverse Osmosis Watermakers

Reverse osmosis is a filtration system for removing unwanted contaminants from water. Like evaporation, it also uses an old and fairly simple technology.

How It Works. To understand how a reverse osmosis watermaker works, you must know some special terms. *Permeability* is a measure of the ease with which molecules flow through a substance. In reverse osmosis, the permeable substance is usually a thin, pliable sheet called a membrane. If the membrane is permeable, molecules can flow through openings in it. But some membranes are *semipermeable*. A semipermeable membrane allows small molecules to flow through it but not large ones. For example, when salt water and fresh water are separated by a semipermeable membrane (fig. 27a), the smaller fresh water molecules pass through the membrane to the saltwater side.

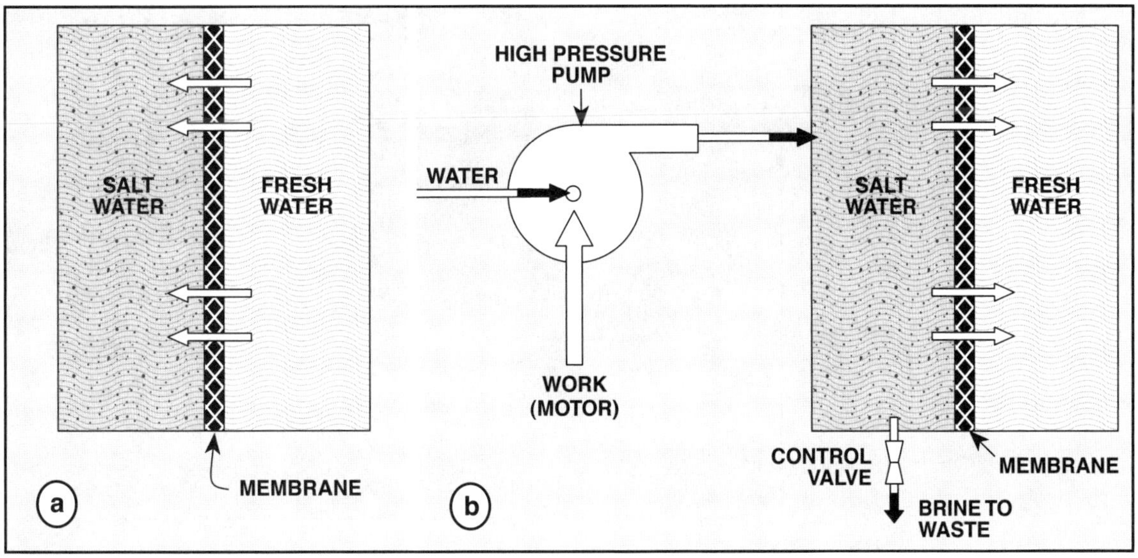

Figure 27a. Osmosis *b. Reverse osmosis*

This passing of molecules is termed *osmosis*. In other words, osmosis is the movement of smaller molecules on one side of a container through a semipermeable membrane to the other side of a container that holds larger molecules. The natural force that causes this migration is called *osmotic pressure*. The process can be reversed (fig. 27b), and fresh water can be made to pass out of the salt solution and into the freshwater side. All it takes is to apply pressure greater than osmotic pressure to the saltwater side—hence the name of the process: reverse osmosis.

Water Flow through the Unit. A reverse osmosis desalination unit (fig. 28) can process salt water to fresh water at a rate of 1,500 to 30,000 gallons (about 6 to 114 cubic metres) per day. Figure 29 shows a flow chart of how the unit works. The supply water, usually seawater, is first chemically treated, as needed, to prevent scale, to flocculate suspended particles, or to remove chlorine. Then it passes through primary and secondary filters to remove particles that could clog the reverse osmosis membranes. The treated water is then pumped by a positive displacement or centrifugal pump to the permeator, the reverse osmosis chamber. The pump provides the pressure for the process to work. Like a mud pump, units with a reciprocating pump have a pulsation dampener to smooth out surges created by the pump as its pistons move back and forth. Minimizing pump surges protects piping and control instruments.

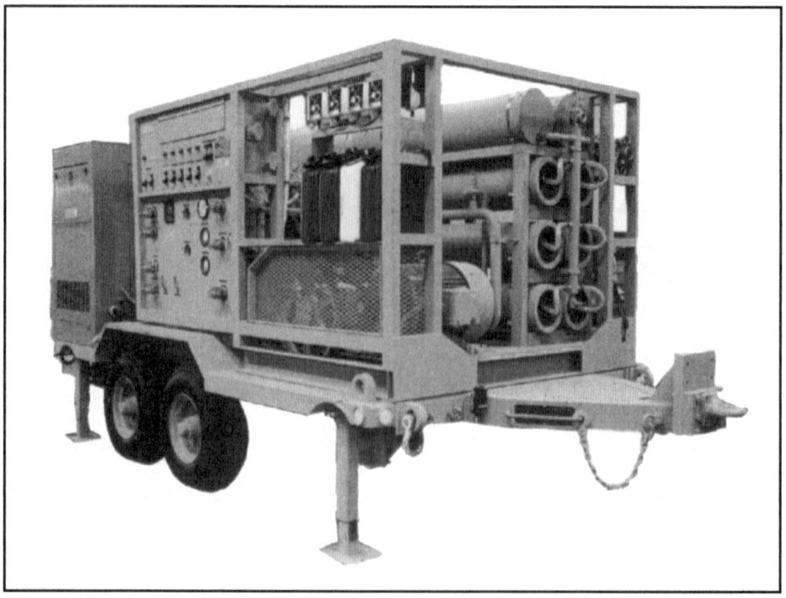

Figure 28. A reverse osmosis watermaker uses a pump to draw salt water through a filtering membrane.

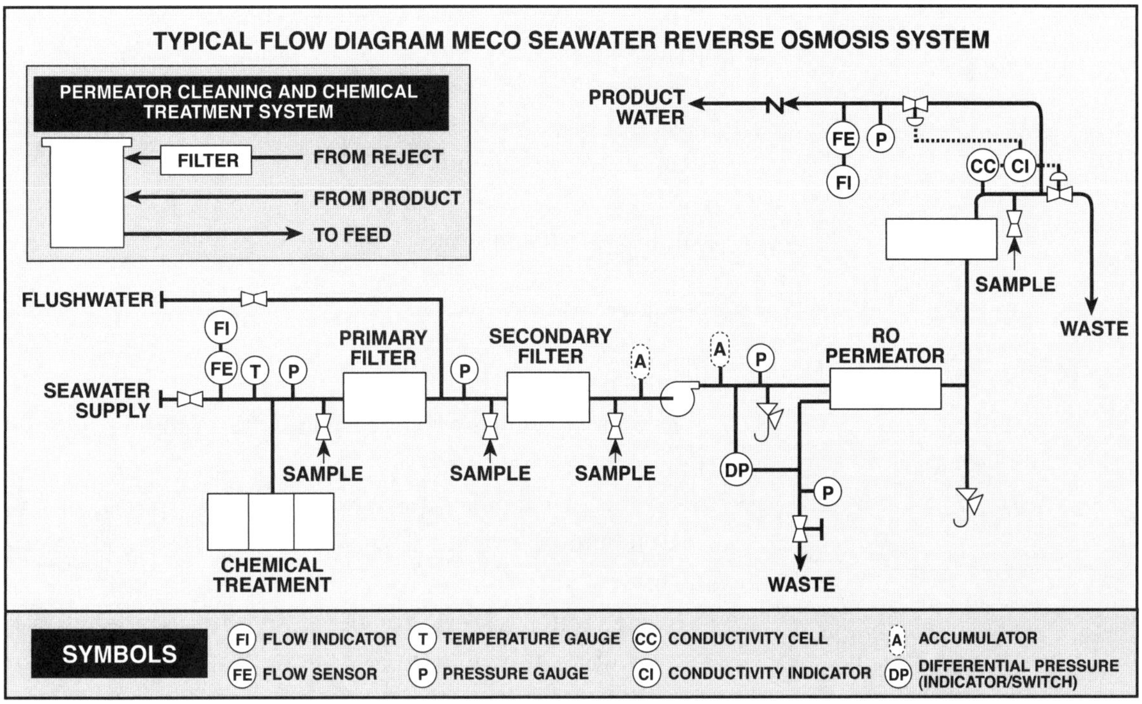

Figure 29. Flow of fluids in a reverse osmosis watermaker.

Depending on the type of membrane, the unit may have a drawback tank that stores fresh water. This water automatically flows back into the permeator to protect the membranes from dehydration after the unit is shut down. Waste water leaves the permeator, while fresh water goes to storage.

Controls. An instrument panel on one end of the unit shows data from flowmeters, a conductivity monitor, an hourmeter, and control relays. The unit may have pressure gauges and other controls as well. A setup for measuring the amount of silt, or fine particles, in the water at different locations in the system may also be provided. An electric control panel houses starters for the motors, circuit breakers, a transformer, fuses, and control wiring.

Alarms. Flashing lights and audible alarms are provided to indicate when the unit is operating outside safe limits, and the system may automatically shut down if the situation is dangerous. Alarms may sound or flash when—

1. pressure across the permeators is too high;
2. feedwater temperature is too high or too low;
3. pump discharge pressure is too high; and
4. conductivity of the product water is too high.

Maintenance. The reverse osmosis unit has a built-in system to clean the permeators. It consists of a mixing tank, a filter, and piping and valves that flush fresh water through the pump, piping, and chambers after shutdown, so that seawater is not left sitting in the unit. If left sitting in the unit for a fairly long period, seawater is very corrosive to the unit's stainless steel parts. The pumps need the usual inspection and lubrication.

To summarize—

- A rig's fuel system uses natural gas, LPG, gasoline, or diesel oil.
- A rig has two auxiliary power systems, compressed air and hydraulic power, that run many auxiliary tools and equipment.
- The rig's water system provides fresh water to cooling systems and for mixing mud and cement, and sometimes for purification for human consumption.
- Watermakers desalinate seawater for offshore drilling and production rigs. The two types of watermakers are evaporators and reverse osmosis units.
- Vapor compression evaporators, waste heat evaporators, and flash watermakers work on the principle that boiling salt water and then condensing the vapor produces fresh water.
- Reverse osmosis units use pressure to draw water through a membrane that blocks the salt.
- The main maintenance problem for most evaporators is cleaning scale from the evaporation surfaces. For reverse osmosis units, the main problem is corrosion due to seawater left inside the unit. Both have pumps that require regular lubrication and inspection.

Rig Cleanup Equipment

▼
▼
▼

Cleaning dirt, oil, and other liquids from the rig site is important for safety and for the environment. A pressure washer or steam cleaner cleans oil and dirt from rig equipment, and a special vacuum cleans up spilled liquids.

Cleaning cuttings is also a consideration on rigs that use oil-based drilling mud. Environmental regulations usually do not allow the rig operator to dump them without washing them first.

Pressure Washers

Pressure washers (fig. 30) use a high-pressure spray of water to clean anything on the rig that is oily or dirty. If a cleaning agent is required, the units provide a place to install a bottle of an environmentally safe detergent to mix with the water. The crew may also use a pressure washer to clean shaker screens, using water or base oil, depending on the type of drilling mud in use. Pressure washers are pneumatic—an air motor powers a triplex pump to pump the water or oil through a hose and wand assembly.

Figure 30. This portable pressure washer can use water, base oil, or solvents to clean equipment.

Maintenance

As is the case with most rig equipment, lubrication is the most important maintenance job. Check the oil levels in the air motor and pump daily, and change the oil on a regular schedule, as the manufacturer recommends.

If the air motor becomes sluggish—that is, begins running slowly or erratically, flush it with a noncombustible solvent. The solvent should flush out any dirt and debris inside the motor. In addition, check the air line water separator regularly and drain off any water that has accumulated. Some units may drain automatically. Periodically, check the pressure gauge to be sure it is working properly so that the unit does not exceed the maximum operating pressure.

Before using the unit, check that the supply and discharge lines from the pump are not blocked, and that the supply line filter is clean. Prime the pump before turning it on—that is, fill the pump inlet with water or base oil so that when it begins running, it is full or nearly full of the liquid in use. Because water or base oil cools and lubricates the pump parts, this action ensures that the pump will not burn out. Wipe the unit down after each use, and flush it out by running water for several minutes to clean out detergent or base oil. If the weather is freezing, empty the pump after each use by running it without water for a maximum of 10 to 15 seconds. Periodically, an experienced mechanic should check the pump valves, pistons, springs, bearings, and seals for wear and cracks.

Steam Cleaners

A steam cleaner sprays pressurized, heated water for cleaning. It may be a portable unit on wheels or permanently mounted on the site (fig. 31). Like a pressure washer, it may also use a detergent additive.

How It Works

The steam cleaner heats water as well as creating a high-pressure spray. The water heater is oil-fired, using diesel, kerosene, or home heating oil as the fuel to heat a coil that extends into a water tank. A triplex pump pumps the heated water through a hose and wand assembly, as in the pressure washer. The pump is powered by a gasoline or diesel engine or an electric motor. A trigger control on the wand allows the operator to adjust the pressure from 500 to 2,000 or more psi (3.4 to 13.8 MPa). A pressure-relief valve and automatic temperature controller ensure safety.

Maintenance

Before starting up the cleaner, check that the power supply cable insulation is in good condition, all hose connections are tight, and the lube oil level in the pump is full. Prime the pump and turn on the unit. Pump maintenance is the same for a steam cleaner as for a pressure washer.

Figure 31. A steam cleaner may be portable or permanently installed at the rig site, such as this one.

Treating Cuttings

A rig that uses an oil-base drilling mud generates cuttings that are contaminated with chemicals and oil. One system for cleaning cuttings is a low-temperature desorption system (fig. 32).

Figure 32. A low-temperature desorption system cleans cuttings before disposal.

How It Works

Workers load the cuttings onto trays to a depth of about 12 to 14 inches (30.5 to 35.5 centimetres) and rake them smooth (fig. 33). A crewmember operating a forklift places the trays inside the treatment chamber. The heaters then slide into place over the trays and are ignited. They raise the temperature of the top few inches (centimetres) of the cuttings to 200° to 500°F (93° to 260°C). A vacuum pump draws hot air from above the trays and through the cuttings to transfer the heat throughout the cuttings. The heat causes the contaminants to boil or vaporize. The vacuum inside the treatment chamber draws the contaminants out through a filter to remove solid particles. The flow then goes to a condenser, which cools the contaminants back to liquid form for disposal. After a certain treatment time, a worker unloads the decontaminated cuttings for disposal. The fuel source for this system may be diesel, propane, or natural gas.

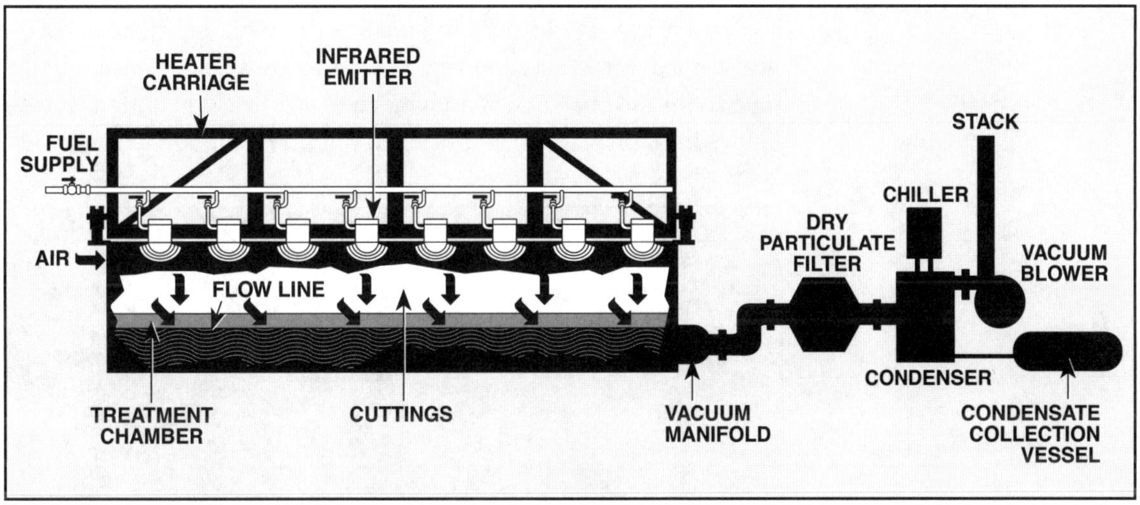

Figure 33. Diagram of low-temperature desorption system

Waste Management

Particularly offshore, waste disposal is expensive and environmentally hazardous. Remembering four steps can help the offshore operator remain profitable and in business:

Reduce waste whenever possible.
Recover consumables that can be reused.
Reuse anything you can.
Recycle items that can be refurbished.

Used lubrication oil is a major disposal problem. To minimize the amount of waste oil, follow this lube oil analysis program:

Analyze fuel.
Select the proper type of engine oil.
Use proper lubrication procedures.
Use the minimum oil necessary.
Document the lubrication schedule.
Vacuum spills.

Drilling mud spilled on the rig floor is one type of waste, and is also a safety hazard. To clean up and recover this mud, the rig may have a heavy-duty vacuum cleaner (fig. 34). Other uses include cleaning the mud room and pipe racks, removing the dregs from tanks, transferring liquids from one tank to another, and skimming the surface of liquids.

Figure 34. The vacuum cleaner picks up spilled liquids.

How It Works

The vacuum cleaner is made up mainly of a vacuum pump, a tank, and hoses. The vacuum pump is air-powered—it uses the rig's compressed air flowing through venturi tubes to create the suction, so no electricity is necessary. (When a fluid like air flows through a device that causes the fluid to bend—change its direction of flow—the change in direction increases the velocity and decreases the pressure. A venturi tube is merely a tube with a tapered constriction in the middle that causes the increase in flow and the decrease in pressure.) The pump draws liquid into the tank through a suction filter and valve.

A float in the tank activates a switch when the tank is full, which diverts the air from the pump to a discharge hose. The vacuum pump shuts down. The compressed air then discharges the liquid in the tank to a storage tank or back to the active mud system. This action may take 10 to 20 seconds. When the liquid level in the tank falls, it activates a low-level switch, which diverts the air back to the vacuum pump to restart suction. The equipment operator controls the pressure of the suction and discharge by setting regulators.

Maintenance

The air supply to the vacuum unit must be clean and dry to prevent the valves and switches from becoming blocked. Drain any liquid from the four air intake filters inside the unit before and after each cleaning job. Remove and empty the suction filter basket before and after each job and whenever the unit loses suction power. If the spilled liquid has a lot of solids in it, dilute it with water or base oil before vacuuming.

Do not use the unit to clean up dry powders or granules unless you dilute them with a lot of water first. At the end of each cleaning job, empty the tank completely by switching the drain-normal switch to drain. Setting the switch to drain causes the unit to empty the tank and hoses, which are then easier to coil for storage.

Because the floats are always in contact with the liquids being recovered, deposits build up on them over time. They then get heavier and less buoyant, causing the unit to behave erratically. One sign of erratic behavior is liquid backing up into the pump. If liquid backs up, drain the tank, remove the floats, and clean them. Cleaning the floats after each job is a good maintenance habit. The twin-action relief valve is on top of the tank under the vacuum pump. Inspect and clean this valve at least once a month.

To summarize—

- Pressure washers use air pressure combined with pumped water to produce a high-pressure stream of water or base oil to clean dirty or oily equipment or shaker screens. Maintenance is mostly lubrication of the pump and air motor.
- Steam cleaners spray heated, pressurized water for cleaning dirty or oily equipment. Pump lubrication and inspection are the main maintenance jobs.
- Cuttings from drilling with an oil-based mud can be cleaned before disposal with a low-temperature desorption system. This system draws hot air through the cuttings to vaporize the contaminants, which are then condensed back to liquid form for disposal.
- Particularly offshore, reducing the amount of waste, especially lubrication oil, is every crew member's business. Reduce, Recover, Reuse, and Recycle.
- A vacuum unit vacuums up liquid spills on the rig floor, mud room, and pipe racks, removes the dregs from tanks, transfers liquids from one tank to another, and can skim the surface of liquids. Maintenance includes emptying the filter basket of solids and draining the unit after using it, as well as cleaning the floats and valves.

Fire Detection and Suppression

Flammable materials are all over a drilling site—oil and grease, natural gas, solvents, rubber hoses, cloth, and paper. Ignition sources are common as well—lit cigarettes, welding torches, and sparks from motors, for example. So fire prevention, detection, and suppression are crucial to safe operation of a drilling rig.

Everyone on a drilling rig should have training in fire prevention and take every precaution to prevent fires—where you see a no smoking sign, for instance, don't smoke. Anyone servicing or operating equipment that involves sparks or flames must know when and how to work safely.

All persons on a drilling rig should know what to do if they see a fire, and know exactly what to do and where to go when a fire alarm sounds. Everyone should know where the rig's fire extinguishers are and how to operate them. Especially offshore, every crew member depends on each other for safety in the event of a fire.

Life Cycle of a Fire

Fire is a form of *oxidation*, a chemical process in which some substance combines with oxygen. The rusting of iron and the rotting of wood are forms of slow oxidation. Fire, or combustion, is rapid oxidation; that is, the burning substance combines with oxygen very quickly. Combustion gives off energy in the form of heat and light.

The Start of a Fire

All matter exists in one of three states—solid, liquid, or gas (vapor). The atoms or molecules of a solid are packed closely together, and those of a liquid are packed loosely. The molecules of a vapor are not packed at all; they move about freely. For a substance to burn, its molecules must be pretty well surrounded by oxygen molecules. The molecules of solids and liquids are too tightly packed to be surrounded. Therefore, only vapors can burn.

When a solid or liquid is heated, its molecules start to move rapidly. With enough heat, some molecules break away from the surface and form a vapor just above the surface. This vapor can now mix with oxygen in the air. If the temperature is hot enough to raise the vapor to its *ignition temperature*—the temperature at which a particular vapor burns—and if enough oxygen is present, the vapor burns.

Burning

Burning is the rapid oxidation of millions of vapor molecules. The molecules oxidize by breaking apart into individual atoms and recombining with oxygen into new molecules (fig. 35). This chemical process gives off energy in the form of heat and light.

The heat from a fire is *radiant heat*. Radiant heat is the same sort of energy that the sun gives off and that we feel as heat. It radiates, or travels, in all directions, so part of it moves back to the burning solid or liquid (the fuel). The portion of the heat that moves back into the burning fuel is called radiation feedback (fig. 36). Radiation feedback causes the fuel to release more vapor and it raises the temperature of the vapor to its ignition temperature. At the same time, air, and thus oxygen, is drawn into the area where the flames and vapor meet. The result is that the newly formed vapor begins to burn, and the flames increase.

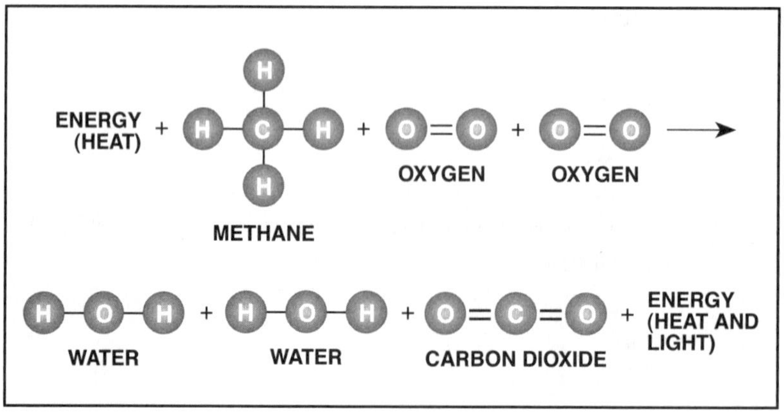

Figure 35. Molecules of a hydrocarbon, such as methane, combine with oxygen when heated. The reaction produces energy we can see and feel in the form of light and heat.

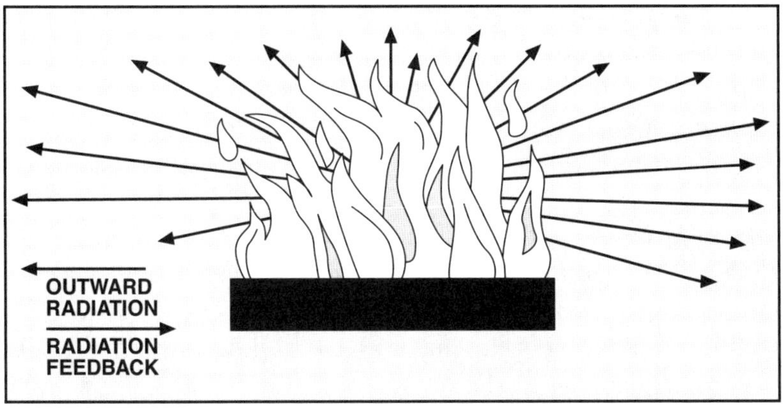

Figure 36. Radiation feedback is heat that travels back to the fuel from the flames.

Growing and Fading

Radiation feedback begins a chain reaction—the burning vapor produces heat, which releases and ignites more vapor. The additional vapor burns, producing more heat and releasing and igniting still more vapor, and so on (fig. 37). As long as fuel is available, the fire continues to grow and produce more flames.

After a time, the amount of vapor released from the fuel reaches a maximum rate and begins to level off. The fire burns steadily and continues to burn until most of the fuel is gone. At this point, less vapor is available to oxidize, and so less heat is produced.

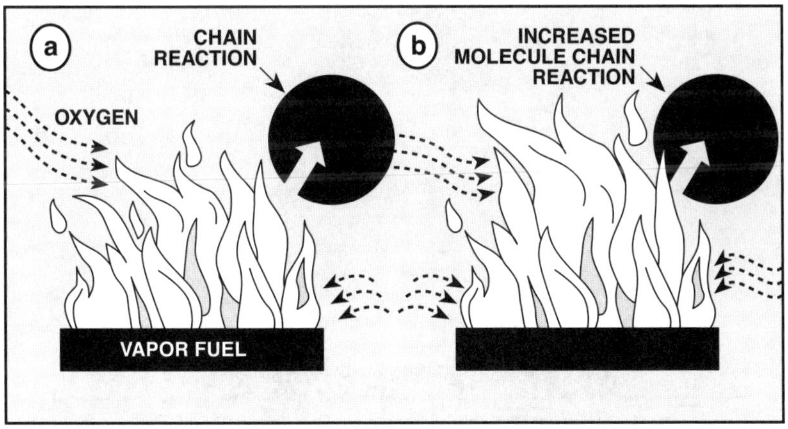

Figure 37. The chain reaction of burning. a) Vapor from heated fuel rises, mixes with air, and burns, producing enough heat to release more vapor and to draw in air to burn that vapor. b) As more vapor burns, the flames grow, producing more heat, releasing more vapor, and drawing in more air.

The chain reaction starts to break down. Still less vapor is released and less heat produced; less heat and flame, and the fire begins to die out. A solid fuel may leave an ash residue and continue to smolder for some time. A liquid fuel usually burns up completely.

Burning Gases

Gases burn more intensely than solids or liquids because they are already in the vapor state. All the energy of radiation feedback goes into raising the temperature of the vapor and igniting it. Gases burn without smoldering or leaving any residue. The size and intensity of a gas fire depends on the amount of fuel available—for example, as a flow from a gas pipe.

Fire Triangle

Combustion requires three things: fuel, oxygen, and heat. The fire triangle (fig. 38) illustrates these requirements. It also illustrates two important rules of preventing and extinguishing fires:

1. If any side of the fire triangle is missing, a fire cannot start.
2. Remove any side of the fire triangle from an ongoing fire, and the fire will go out.

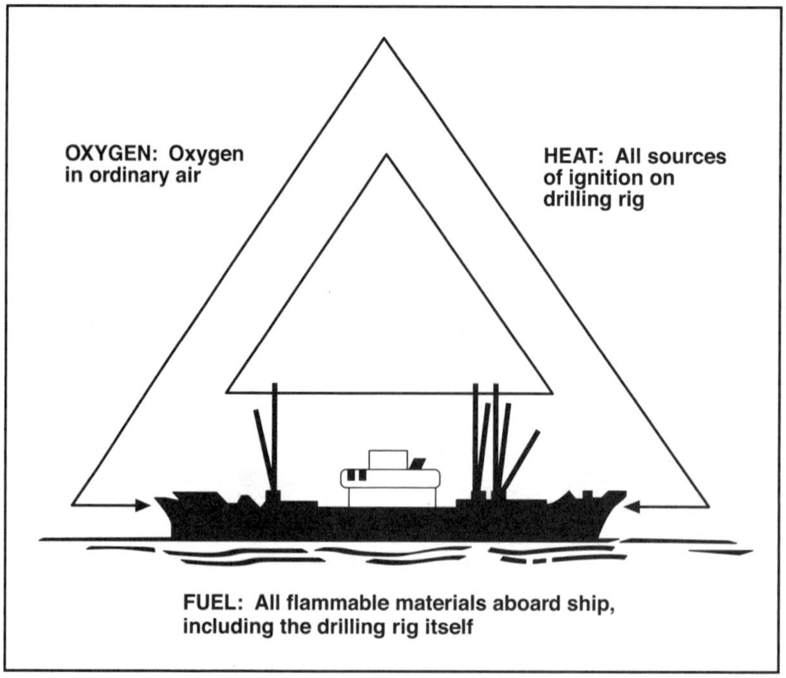

Figure 38. The fire triangle. Fuel, oxygen, and heat are necessary for combustion.

Extinguishing a Fire

The purpose of fire suppression equipment is to remove the heat, the fuel, or the oxygen from a fire, or to interrupt the chain reaction of burning.

Removing the Heat

The most common way to extinguish a fire is to remove the heat with water. Water absorbs heat very well, both from the fuel and from the radiation feedback. Drilling rigs on land and offshore have hoses for directing water to a fire.

Removing the Fuel

One way to remove the fuel is to physically drag it away. With a solid fuel, this is usually impractical, although removing nearby fuel that is not yet burning is a good way to prevent the fire from spreading.

When the fuel is a liquid or gas coming from a line, shutting off the proper valve will remove the fuel source. Some equipment has safety valves that can be shut off by the pressure from a stream of water.

Removing the Oxygen

Removing or reducing the oxygen puts out a fire by smothering it. One way to smother a fire is to flood the area of the fire with an inert gas, such as carbon dioxide, which displaces the oxygen. Inert gases do not react with either the fuel or oxygen. This method is difficult or impossible to use in an open area. Carbon dioxide (CO_2 for short) would quickly blow away from an open deck, for example. It is useful for fires in contained areas, such as a sealed compartment or room.

Another method of removing oxygen is to smother the fire with foam. Foam floats on top of a burning liquid, for example, and spreads out quickly to form a blanket over it. This blanket cuts off oxygen to the fire, as well as cooling it down. The hose setups that supply water for fire suppression often have a small hose that can feed a foaming agent into the water stream.

The U.S. Air Force has developed a new class of powerful fire suppression agents known as encapsulated micron aerosol agents, or EMAA. EMAA is stored in the form of a solid, powder, or gel. When it is ignited, it forms an *aerosol*, a suspension of very small particles in a gas. Smoke and fog are types of aerosols. The aerosol is lighter than air and smothers fires in enclosed areas. EMAA is also effective against fuel tank fires. The Air Force has put it at the bottom of a fuel storage tank and then set it off when the fuel catches fire. The resulting aerosol percolates to the surface of the fuel and quickly extinguishes the fire. EMAA is undergoing tests for environmental and occupational safety.

Breaking the Chain Reaction

Some dry chemicals and hydrogenated hydrocarbon gases, called Halons, extinguish a fire by interrupting the chemical reaction of the fuel with oxygen, which stops the flames. This method is effective on liquid and gaseous fuels because they must flame to burn. Although Halons are very effective fire extinguishing agents for certain types of fires, they are illegal in most drilling areas for environmental reasons. Only facilities that had Halon systems already in place before the gas was outlawed can use it. Replacements are now available, but so far they are two to three times less effective than Halons.

Classifying Fires

Choosing which method to use to extinguish any particular fire relies on a system for classifying fires developed by the National Fire Protection Association. The system classifies fires into four types, according to the type of flammable material.

- *Class A fires:* Fires that involve common ash-producing materials, such as wood, paper, cloth, rubber, and some plastics (fig. 39). Water and foams are the best extinguishers for a Class A fire.

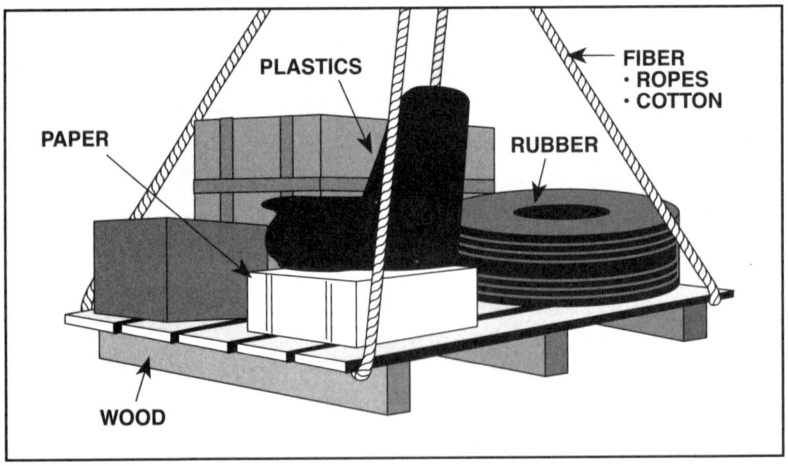

Figure 39. Class A fires involve common flammable materials.

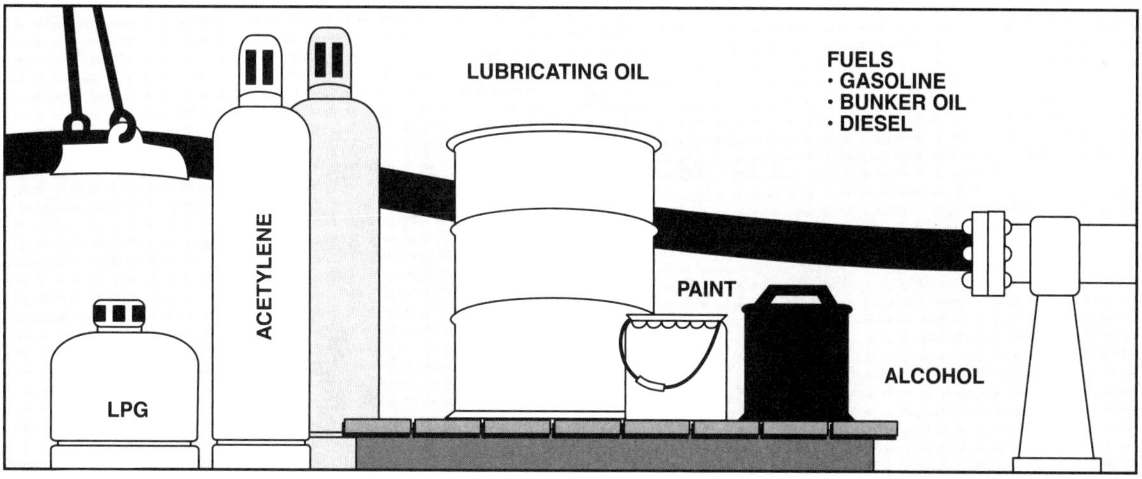

Figure 40. Class B fires involve flammable liquids, gases, and petroleum products.

- *Class B fires:* Fires that involve flammable liquids, gases, and greases, including petroleum and natural gas, as well as alcohols and paints (fig. 40). The best extinguishing method for Class B fires is smothering them to cut off the supply of oxygen.
- *Class C fires:* Fires that involve energized electrical equipment and lines (fig. 41). The extinguisher for these fires must *not* conduct electricity, so water is out. The best choice is to use CO_2 or a dry chemical to smother the fire or break the chain reaction.

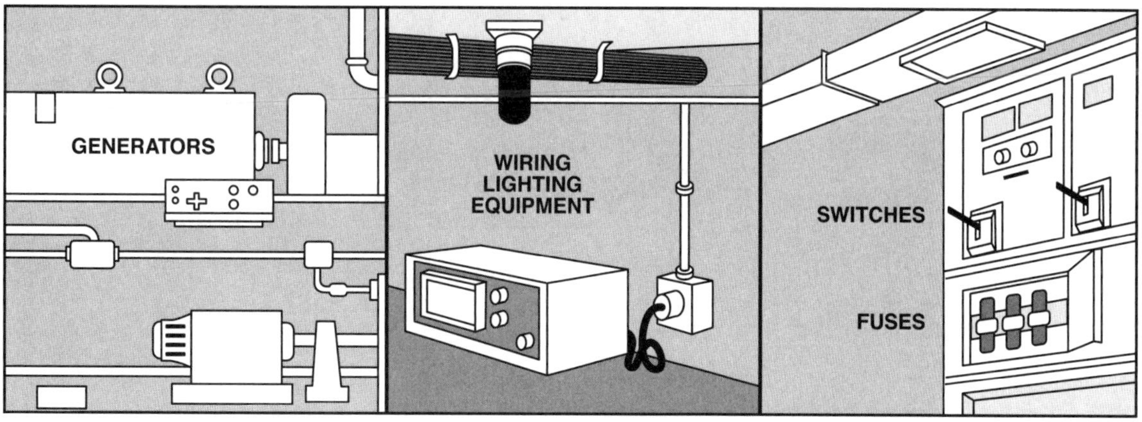

Figure 41. Class C fires involve live electrical equipment and wiring.

- *Class D fires:* Fires involving combustible metals, such as zinc, titanium, aluminum, and sodium. These metals are particularly hazardous when in powdered form, where they can become airborne and explode. They burn at such a high temperature that water does not work as a cooling method (fig. 42). Special extinguishing agents are available for each metal to smother the fire. Fortunately, Class D fires are rare on the drilling site.

In reality, some fires require the use of more than one extinguishing method. Firefighters extinguishing a Class B fire, which includes burning oil storage tanks and flow lines, for instance, use both water for cooling the metal and foam for smothering the flames.

Figure 42. Class D fires can become airborne and explosive when in powdered form.

Fire Suppression Equipment

Fire suppression equipment on the drilling site includes portable and semiportable fire extinguishers and fixed systems for delivering water, foam, inert gases, or chemicals to the fire.

Portable Fire Extinguishers

Portable extinguishers may be small enough to carry (fig. 43) or large enough to require a wheeled caddy (fig. 44). They are useful for quickly attacking a small fire. Even the larger extinguishers carry only a limited amount of extinguishing agent which is expelled very quickly, in 20 to 60 seconds.

Figure 42. A hand-held fire extinguisher has a short range of 6 to 8 feet (1.8 to 2.4 metres).

Figure 43. A wheeled extinguisher has a hose for a longer range.

Labeling

Extinguishers are clearly marked as to the class of fire they will put out: A, B, or C (fig. 45). Some are rated for more than one class—AB, BC, or ABC. Do not use an extinguisher on a fire for which it is not rated.

If the extinguisher has an A or B classification, it also has a number marking that indicates its size or efficiency. The higher the number, the more fire the extinguisher can put out. For example, an extinguisher rated 4A will extinguish twice as much Class A fire as one rated 2A. A 20B-rated extinguisher will extinguish four times as much Class B fire as one rated 4B. Class A extinguishers have number ratings from 1A to 40A; Class B from 1B to 640B.

A Class C extinguisher has no number rating because this classification is simply added to a Class A or B rating to indicate that the extinguisher is safe to use on a fire involving electrical equipment.

A Class D extinguisher is rated only for a particular combustible metal, and is never suitable for use on a Class A, B, or C fire.

The Coast Guard uses a Roman numeral instead of an Arabic number to indicate the extinguisher's size. The numeral I is the smallest size, and V is the largest. For example, a BIII rating means that the extinguisher is medium sized and suitable for Class B fires.

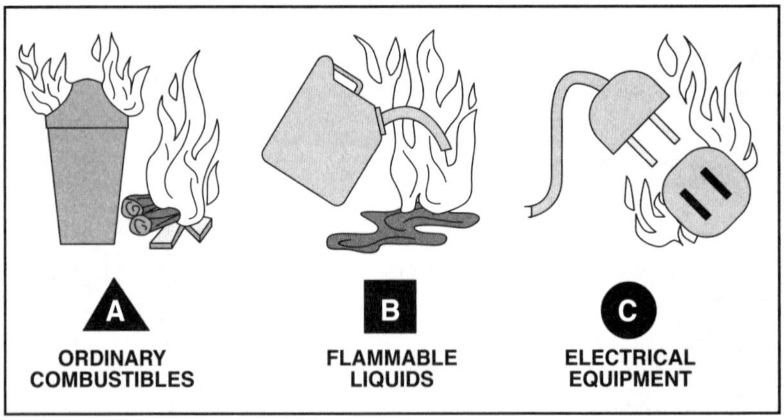

Figure 45. Fire extinguishers are labeled with a letter inside a geometric shape for their class (A, B, or C) as well as a picture symbol for the type of fires they put out.

How to Use a Portable Extinguisher

The extinguisher's label should also have detailed directions on how to use it. In general, for a hand-held type, hold the extinguisher upright, pull the pin on top that keeps the handle from being pressed, aim the nozzle or short hose toward the fire, and squeeze the handle. Sweep the nozzle back and forth at the base of the flames.

Wheeled types work the same way—after wheeling the extinguisher to the fire, unravel the hose, and spray at the base of the flames.

Extinguishing Agents

The agent inside the extinguisher varies with the classification. Class A extinguishers may contain water or a water solution, foam, or dry chemical. Class B extinguishers contain foam, dry chemical, or CO_2. The agent in a Class D extinguisher is a dry powder (not dry chemical).

Many portable extinguishers use a cartridge (fig. 46) of nitrogen or carbon dioxide to supply pressure to expel the water, foam, or chemical inside. This cartridge is replaceable and the extinguishing agent can be refilled.

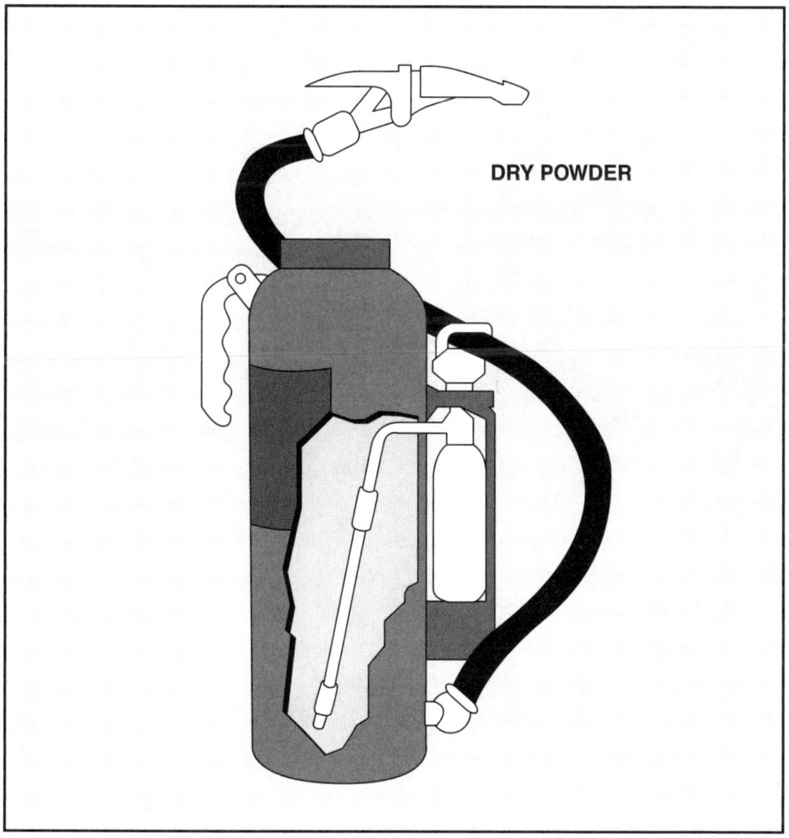

Figure 46. A separate cartridge filled with pressurized nitrogen or carbon dioxide expels dry chemical extinguishing agent.

Inspection

To be sure that a portable extinguisher will be in working order when it is needed, make the following checks at least once per month:

A. Check that the extinguisher is in the proper location and is accessible.

B. Make sure the operating instructions and labeling are readable.

C. Inspect the nozzle for obstructions, such as an insect nest.

D. Look at the hose and fittings for corrosion or cracks.

E. Check the lock pin to be sure that no one has tampered with the extinguisher.

F. Make sure the extinguisher is full by weighing it or inspecting the level. Also check the pressure by looking at the pressure gauge. If it is not full and fully pressurized, replace it or recharge it.

G. Check the inspection tag with your name and the date of the inspection.

Some drilling operators hire a fire protection company to make these inspections, but they are not the people who will be burned if a fire gets out of control due to a faulty or damaged extinguisher.

Fixed Systems

Fixed fire suppression systems are permanently installed at certain locations on the drilling rig. The category includes several different types of delivery systems and extinguishing agents—a fire-main system for supplying water, foam systems, CO_2 systems for flooding enclosed spaces, and automatic sprinklers. Fixed systems are the big guns of fire suppression on a drill site.

Fire-Main System

A fire-main system is a system of piping that carries water from pumps to several locations on a rig called fire stations. Water's fire suppression ability comes from its cooling effect. At each fire station is a water outlet and a monitor (a fixed nozzle setup) and/or hoses with nozzles. The hose or monitor stream from one fire station overlaps the stream from the adjacent station, so the whole structure is protected.

In warm climates, the piping is filled with water. In cold climates, it is either dry or filled with a mixture of water and antifreeze. Two diesel-powered fire pumps (a primary pump and a backup) supply water to the system at pressures and volumes high enough to provide a strong hose stream. On offshore rigs, seawater is the water source. On land rigs, the water may come from a neighboring lake or pond, a city water supply, or a water well. In an emergency, the crew may even stop passing trucks carrying brine out from nearby drilling operations.

Monitors

A monitor is a heavy-duty metal structure with a nozzle attached that sits on the floor or deck. Its purpose is to allow one person to control a high-gallon water flow. A portable monitor sits on three legs and receives water through a hose (fig. 47a). A fixed monitor is bolted to a water supply pipe (fig. 47b). Because high water pressure and volume through a portable monitor can cause it to slide or get out of control, it has a valve that reduces the water flow automatically if the monitor loses contact with the ground.

Monitors swivel and tilt up and down by means of a handle and an elevation control knob. They can deliver 500 to 2,000 gallons (1,892 to 7,570 litres) of water per minute in a solid or fog stream. A stream that size can extinguish a large fire and also cool down metal supports and decks or floors. The advantage of a monitor is that it can attack a fire from a relatively safe distance, making it especially useful on helicopter pads on offshore ships and platforms. In the event of a helicopter accident, a monitor can blanket the whole landing pad with water quickly. The water either extinguishes the fire or keeps it under control until firefighters can come in with dry chemical extinguishers or hoses.

The number, capacity, and location of monitors are calculated as a part of the total design of a fire suppression system. Helipads always have both monitors and hand-held hoses, and other monitors generally are located near production and product storage areas.

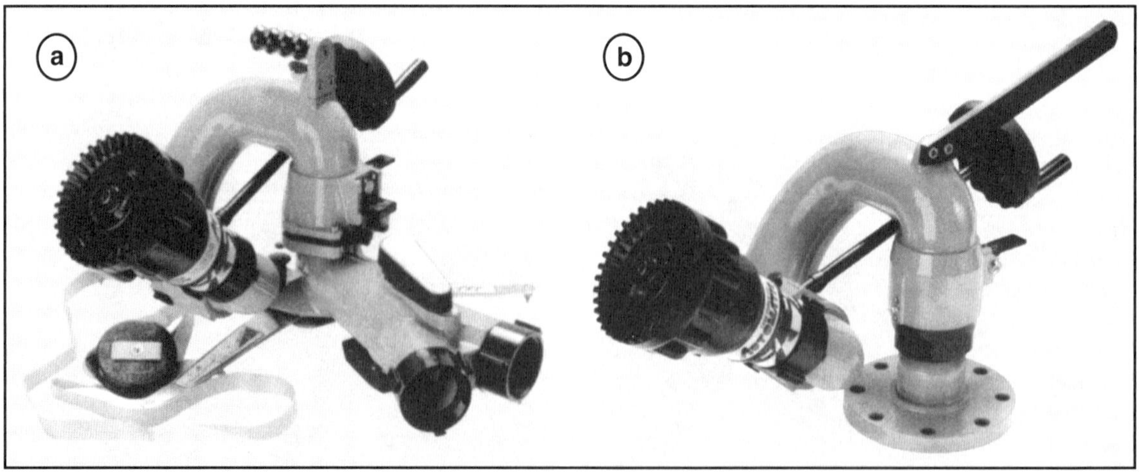

Figure 47. A monitor sits on the ground (a) or is permanently bolted to the floor or deck of a drilling unit (b).

Hoses

The size of a hose determines the amount of water that can flow through it, or *gallonage*. And the gallonage determines how many people it takes to handle the hose. Hoses 1½ or 1¾ inches (38 or 45 millimetres) in diameter are small enough that two people can handle them. Larger hoses (2½ and 3 inches, or 65 and 77 millimetres) require three to five people to hold and direct them when water is flowing. Each hose is 50 feet (15.24 metres) long and has a male coupling on one end to attach the nozzle and a female on the other that screws into the water hydrant. Some are lined to strengthen the hose (fig. 48).

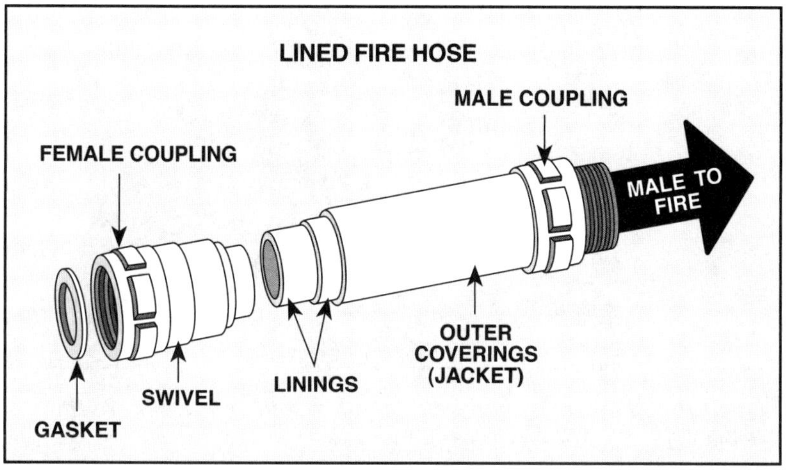

Figure 48. Lined fire hose and hose couplings

The smaller-size hoses are *attack hoses*, for moving in towards a smaller fire or one that is more or less under control. The larger sizes are *defensive hoses*, used, like monitors, for standing back from the fire and throwing large amounts of water at it or for supplying water to portable monitors. Defensive hoses are often installed at special fire-fighting points called fire stations (fig. 49).

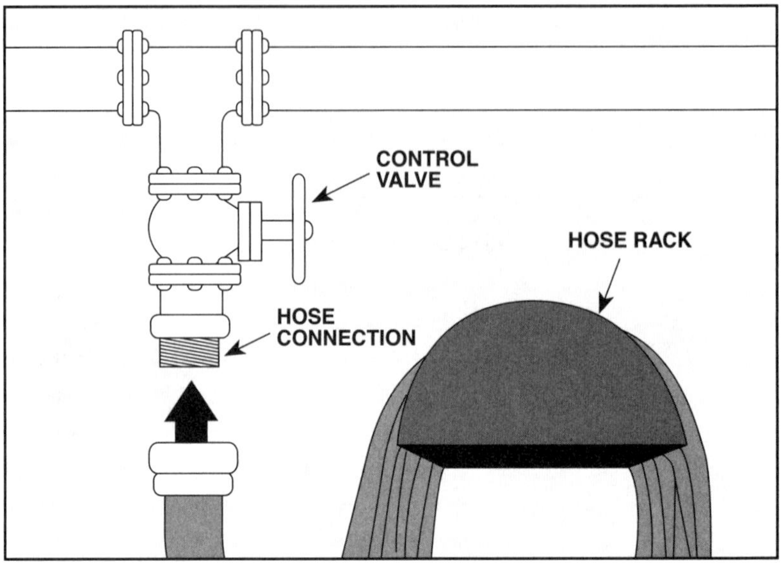

Figure 49. The three components of a fire station hydrant are the control valve, hose connection, and hose rack.

Nozzles

Nozzles can produce two types of sprays—a narrow, solid stream of water or a wide, fine spray called a *fog*. An adjustable fog nozzle can deliver a solid stream or one of two fog patterns (fig. 50). When its U-shaped handle, the *bail*, (fig. 51) is pushed forward, it closes a valve that shuts off the water flow. The farther back the handle is pulled, the more water can flow through. Rotating the barrel of the nozzle changes the stream pattern from solid to 30° fog to 90° fog.

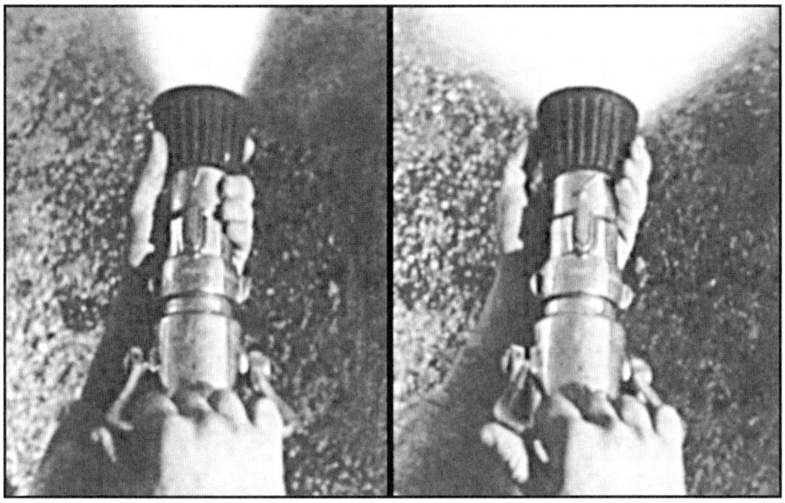

Figure 50. An adjustable fog nozzle delivers a 30° or 90° fog pattern as well as a solid, narrow stream of water.

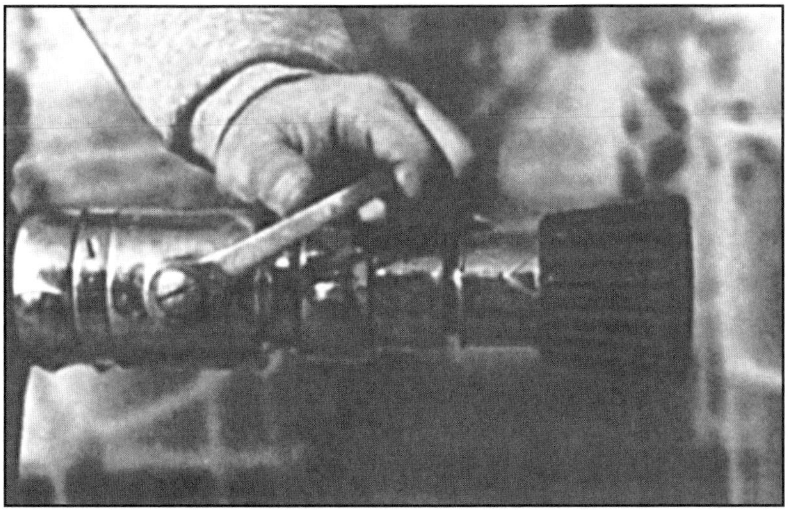

Figure 51. When the bail is pushed all the way forward, the nozzle is closed.

Maintenance

Once a month, check the following:

- Make sure all the parts of the fire-main system are where they belong—nozzles, hoses, portable monitors, tools.
- Check the monitors for leaks, and that the valve handle moves freely between the open and closed positions.
- Make sure the swivels on monitors rotate while water is flowing.
- Lubricate mechanical components as recommended by manufacturers.
- Check that hoses are not kinked and that the threads are free so that the hose can be connected to the nozzle and hydrant by hand.
- Check the gaskets in the nozzles for cracks and wear, and clean out any obstructions inside the nozzle.
- Test the nozzle's bail and adjustment ring for free movement.

Foam System

Each fire station also has a foam station, a tank of foam concentrate with special nozzles and a device that feeds foam concentrate into the water stream. In particular, the helipad will have a foam station as may processing, storage, or handling areas where foam is useful for controlling spills of flammable liquids. The foam system can be portable, carried from one station to another as needed, or it can be a fixed system in its own foam room with large containers of foam concentrate.

Mechanics of the Foam System

The system consists of a container of foam concentrate with a pickup hose that feeds into the water hose. A venturi device of some sort feeds the concentrate into the water stream and a special nozzle adds air to create the foam. As mentioned earlier, a venturi device works on the principle that a fluid flowing through a constriction has increased velocity and reduced pressure. The water passes through the venturi so fast that it lowers the pressure at the opening where the foam pickup hose enters, which creates a vacuum. The vacuum sucks the foam concentrate into the stream of water.

To create foam requires four ingredients: foam concentrate, water, air, and mechanical agitation. The foam nozzle adds air in the same way a simple bubble screen in a detergent solution creates bubbles. A child either blows through a piece of screen or waves it in the air, and the detergent mixture forms bubbles. A foam nozzle has a similar sort of screen that causes the foam-water mixture to become turbulent and take in air.

Some foam systems use a *mechanical nozzle*, where the venturi device and the agitation device are both in the nozzle (fig. 52).

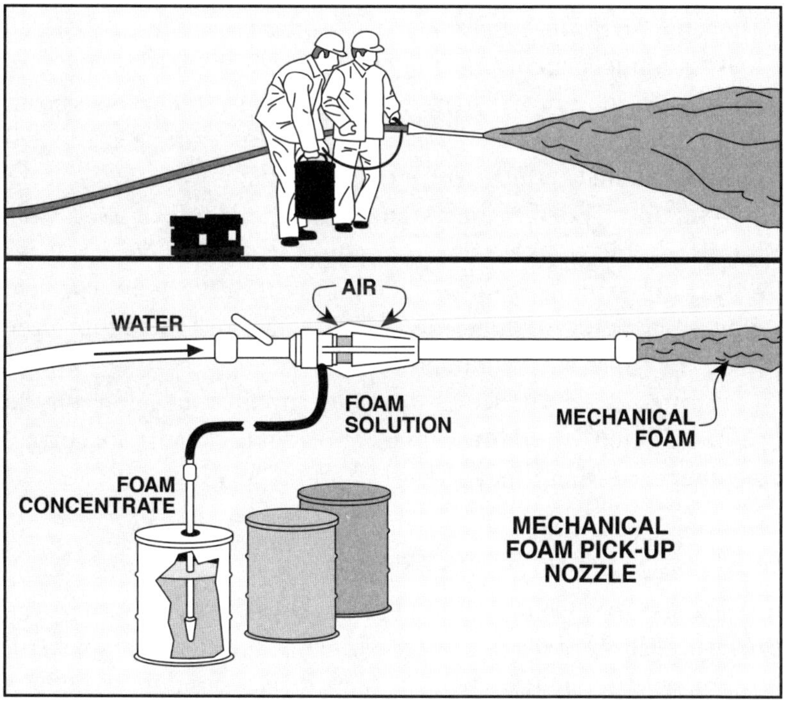

Figure 52. With a mechanical nozzle, the nozzleman's movements are restricted by the need to keep the pickup hose in the foam concentrate container.

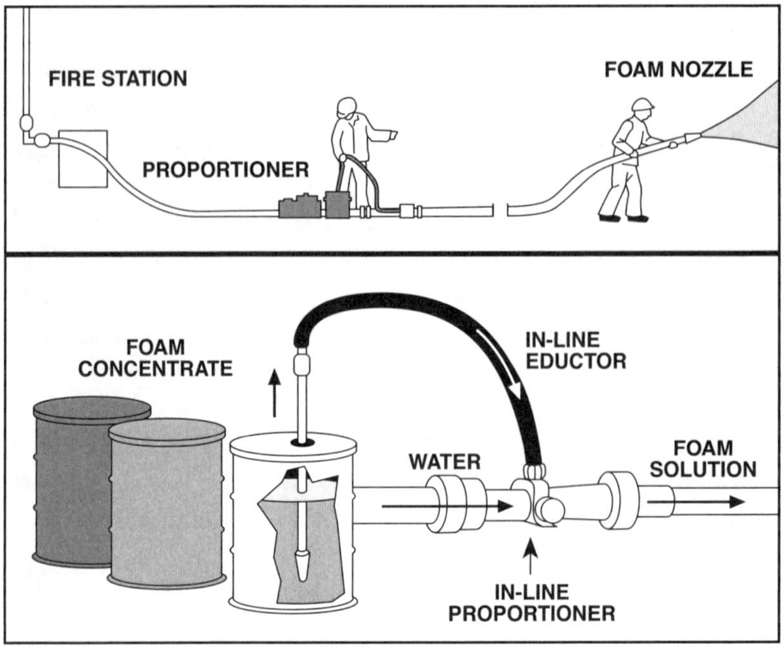

Figure 53. An in-line proportioner allows the foam concentrate container to be farther back from the end of the hose.

Others use an *in-line proportioner* that screws into the ends of two hoses to connect them (fig. 53). The in-line proportioner uses a venturi tube to draw the foam concentrate into the water stream at a predetermined proportion. Then the foam nozzle at the end of the hose agitates the mixture to create foam.

With any type of foam system, extra containers of foam concentrate should be available and open. Depending on the water pressure, a 5-gallon (19-litre) container lasts only about 1½ to 2½ minutes, so a crew member needs to be standing by to move the pickup hose into a new container.

Maintenance for a foam system is essentially the same as for the fire-main system. In addition, make sure that the foam concentrate containers are where they should be and plenty are available.

How a Foam System Suppresses Fire

Foam works in several ways to put out a fire. First, it smothers the fire by preventing oxygen from reaching the fuel. Since foam is still 99 percent water, it also cools the fuel and surrounding structures as well as water does. Finally, foam allows water to penetrate into the fuel and absorb even more heat.

CO_2 System

Storage areas and rooms housing the engines, mud pumps, generators, compressors, and electronic equipment usually have a CO_2 system, which replaces the oxygen in the room with nonflammable CO_2. Before Halons were banned, these areas may have had Halon systems. Most rigs went back to using CO_2, but a few use other gases developed to replace Halons.

The system consists of a group of cylinders full of pressurized carbon dioxide connected to a manifold and connected to each other with pressure hoses. The manifold has a master stop valve which opens to let the CO_2 flow down a network of pipes toward the discharge nozzles located in the protected space. The number of cylinders, pipe size, number of nozzles and their diameter depends on the size of the room and the concentration of carbon dioxide needed to protect it. The entire system operates using pressure, so no external power is necessary.

The CO_2 system may be manual or automatically activated by a fire detector. When activated, a warning alarm sounds and the CO_2 first flows into a pneumatic timer cylinder. This timer delays the release of the gas for 20 to 40 seconds, giving any people in the room time to get out. (Cutting off oxygen to a fire cuts off oxygen for breathing at the same time.) The detector generally closes any exhaust fans and dampers automatically to seal the room so that the CO_2 cannot escape.

Automatic Sprinklers

The living, recreation, office, and galley spaces on a fixed offshore unit may be protected by an automatic sprinkler system. This system consists of piping, sprinkler heads, valves, pumps, and a pressure tank.

When fire causes one or more sprinkler heads to open (fig. 54), the initial water supply comes from the pressure tank. As the water flows out of the tank, the water itself or the pressure drop in the tank causes alarms to sound and starts the fire pump, which supplies more water to the system.

Sprinkler systems are reliable if they are well maintained. When it fails to operate, it is usually because the wrong valves are closed. To avoid this, mark all valves with their operating position (such as "Keep open at all times") and, if necessary, sealed in the proper position.

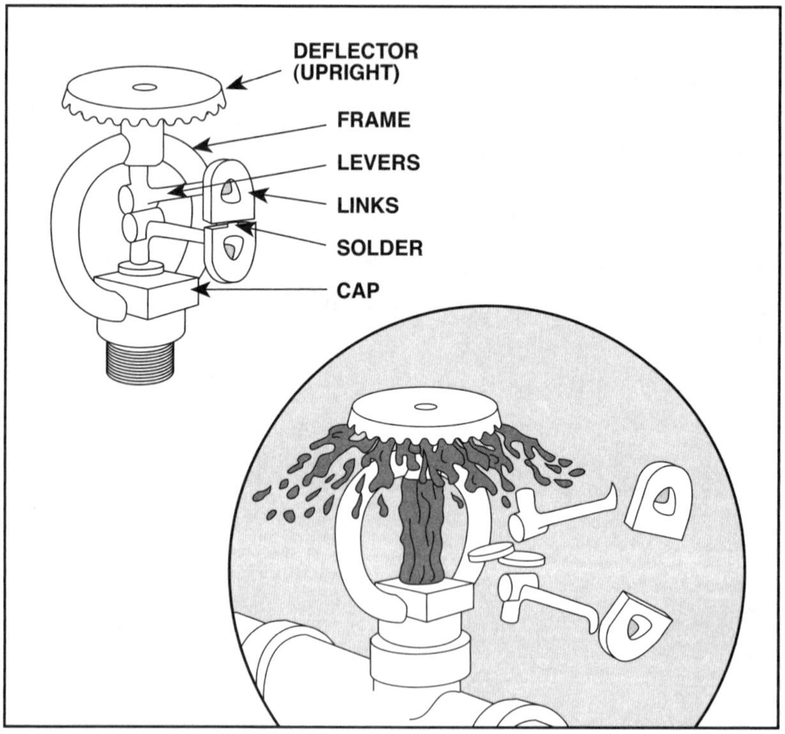

Figure 54. Heat from a fire melts solder, which allows the links to separate. The levers come apart and water pressure pushes the valve cap off the sprinkler outlet. Water then flows up against the deflector, forming a spray that falls onto the fire.

Water Spray Systems

Some offshore rigs use a water spray system to protect the wellhead. The water spray system uses pumps, piping, and valves to send water through spray heads that provide an umbrella of water over the protected area. Other heads direct water to nearby walls or equipment to cool it. This system can be manual or automatic. When a detector activates this system automatically, the fire pumps begin pumping and a *deluge valve* opens. A deluge valve opens fast and allows the full volume of water to flow immediately.

Fire Detection Equipment

Since fire is a continuous threat on drilling rigs, most have automatic fire detection systems. The detector sounds an alarm when it is activated, and, in unattended areas, automatically turns on the fire extinguishing system. The rig also has a manual fire alarm system, and crew members can report fires by telephone and intercoms.

Fire Line Automatic System

Also called a *pneumatic tube fire detector*, this system is the only one that can detect fires in open spaces. It consists of a length of flexible plastic or metal tubing that forms a loop around the outside of whatever structure it is protecting (fig. 55). The tubing is filled with air or gas under pressure. The ends of the tubing are connected to pressure switches. When a fire burns through the tube, the pressure drop causes the pressure switches to flip on. The switches may control alarms and shutoff valves, and may turn off machinery and equipment and turn on fire-main pumps.

The fire line system is simple and reliable, but its drawback is its vulnerability to false alarms. It will activate the equipment it controls if the tubing is accidentally cut or damaged by abrasion.

A rig may have several fire line systems. For example, on an offshore unit, each deck may have its own loop. A separate loop may control the water spray system at the wellhead. Other areas that may have their own loops are the rooms protected by CO_2 systems. Here the tubing is metal with plugs that melt when the temperature reaches a certain point.

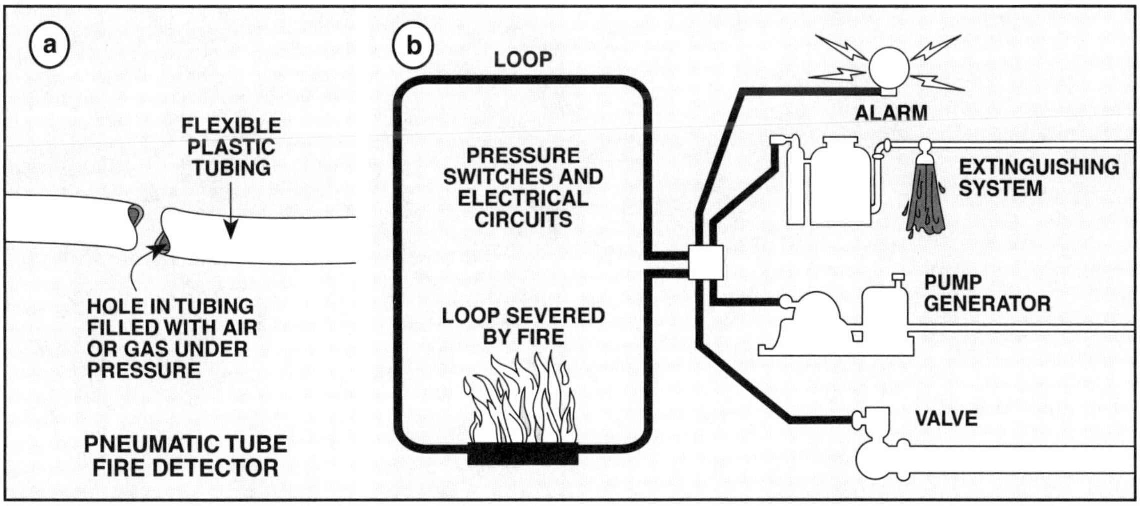

Figure 55. Fire line automatic system. a) The tubing is filled with air or gas under pressure. b) When fire melts the tube, the loss in pressure activates switches that turn on alarms, the extinguishing system, pump generator, and product control valves.

Heat and Smoke Detectors

Heat and smoke detectors are spot detectors, like ordinary household smoke alarms. They are installed mainly in living spaces, rooms housing electronic equipment, and storage areas. These types of fire detectors are useless in open areas, where wind could carry away the heat and smoke of a fire. Usually they only sound an alarm when they sense fire.

Combustible-Gas Detector

Another type of detector used in living areas, galleys, and equipment rooms is the combustible-gas detector (fig. 56). They may also protect product pipelines, manifolds, and the wellhead. This detector warns of dangerous concentrations of combustible, or flammable, gas. While they are not fire detection systems as such, they do detect situations that could lead to an explosion and fire.

One type of combustible-gas detector draws air into a chamber and heats it. Any combustible gases in the air sample burn, which changes the electrical output of a circuit in the unit. A meter shows the concentration of the combustible gas, and when the concentration reaches the danger point, the detector sets off an alarm.

Figure 56. A combustible-gas detector senses the presence of flammable vapors in the surrounding air.

Manual Fire Alarms

No matter how many detectors a rig has, the crew should always watch for fires. Many times an alert crew member discovers a fire before even the most sophisticated detection system. In spaces that have no such device, the crew members are the fire detectors. When someone sees a fire, he or she must go to the nearest manual alarm and sound it immediately. The most common manual alarm system is electrical, powered by the rig's generators, with batteries as a backup power source. The system includes fire alarm boxes (fig. 57) on all levels, in open and enclosed spaces, usually with telephones next to them.

Each alarm box has three buttons—yellow or orange, red, and black. Pushing the yellow or orange button sounds the fire alarm, a warbling two-pitched siren. This sound means that everyone should report to their fire stations. Pushing the red button produces a steady-pitched siren that is the signal to abandon the rig. The black button shuts off the siren, used if the alarm was set off accidentally or in error.

Figure 57. A rig has several manual alarm boxes.

Personal Safety Equipment

The most important fire safety equipment any rig worker can have is thorough training in the prevention of fires, the location and operation of alarms and extinguishers, and when to use them. Those who man hoses and fixed fire suppression systems need additional training. Particularly on an offshore rig, there is no one but the drilling crew to put out a fire and nowhere to go to escape it except the ocean. Weekly fire evacuation drills take place at random times—usually, it seems, while sleeping after working a 12-hour shift.

Concretely speaking, firefighters have access to protective clothing—called a *bunker suit*. A rig usually has 6 to 12 protective suits. Additionally, all rooms have life jackets.

Bunker Suit

The bunker suit (fig. 58) consists of:

- Helmet—protects the head from impact and puncture injuries; the face shield protects the eyes from flying solids or liquids.
- Hood—protects the part of the face, ears, and neck not covered by the helmet or coat from heat.
- Coat and pants or long coat—protects the body against cuts, abrasions, and burns, and provides limited protection against corrosive liquids.
- Gloves—protect the hands from cuts and burns.
- Boots—protect the feet from burns and puncture wounds.
- Air supply—provides oxygen to breathe and protects the lungs and respiratory tract from dangerous gases, heat, and smoke.

Coat and Pants

The coat and pants are made of three layers—an outer shell, a moisture barrier, and a thermal barrier. These layers protect against flame, cold temperatures, and hot water and vapors. If the wearer does not suit up properly, however, he or she will not be completely protected. For example, the coat has a double-layered storm flap closure in front. Be sure to close both the snaps or zipper inside and the Velcro flap outside. Just closing the flap and not the inner closure could allow the coat to fly open. Closing only the inner closure and not the flap allows water and steam to get inside between gaps.

Figure 58. A bunker suit reflects as much as 90% of the radiant heat from a fire.

Air Supply

Because fires consume oxygen, an air supply is crucial to the firefighter. Fires also release carbon monoxide and small particles of carbon, tar, and dust that form smoke. Inhaling carbon monoxide or smoke can quickly disable a person. The air supply also protects the wearer in a room with a CO_2 system that has been discharged, and in areas where burning materials produce dangerous gases, such as hydrogen cyanide and hydrogen chloride.

The air supply consists of a cylinder of compressed air or oxygen, straps to hold it onto the wearer's back, a face mask with hose, and a regulator assembly to regulate air pressure. Cylinders come in different sizes; a typical one is rated for 30 minutes of air and weighs from 9.6 to 23.8 pounds (4.3 to 10.7 kilograms), depending on whether it is aluminum or steel. The regulator reduces the pressure of the air from the cylinder to slightly above atmospheric pressure and controls the flow of air. Inhaling moves a diaphragm out so that air can move in, and exhaling moves the diaphragm back to the closed position. A pressure gauge that shows the pressure of the air remaining in the cylinder is mounted next to the regulator. An alarm sounds when the pressure falls below a certain amount. This means that the air supply is about to give out, and the firefighter should leave the area of the fire immediately.

The face piece is made of clear plastic with a rubber mask that fits snugly against the face by means of straps. The hose brings air from the regulator to the face piece, so be sure to keep it free of kinks. The hose is corrugated to keep it from collapsing when the wearer leans against a hard surface.

Maintenance of Protective Clothing

A label on the coat will give instructions on how to keep it in good shape.

- Wash all oil, grease, chemicals, and dirt from the bunker suit, boots, and face piece after use. Contaminants may cause the suit fabric to deteriorate, and dirt absorbs heat faster than the protective outer layer of the suit.
- Repair tears according to the manufacturer's instructions.
- Store the clothing properly.
- Make sure the air cylinder is full.
- Check all gauges and the low-pressure alarm on the air supply.
- Check the face piece and hose by inhaling with your hand covering the end of the hose.
- Connect the hose to the regulator and inhale to check the regulator.
- Check that all straps are in good condition.

To summarize—

A fire requires three things: fuel, oxygen, and heat. If any of these are absent, a fire cannot start; removing any of these from a fire will extinguish it.

Fire Extinguishers

- Fire extinguishers contain different types of extinguishing agents that work on different classes of fires—A, B, or C. Always match the class of fire to the label on the fire extinguisher.
- Know where fire extinguishers are located, and check regularly that they are full and pressurized and undamaged.

Fixed Fire Suppression Systems

- A fire-main system consists of fire pumps that pump water through pipes to several fire stations around a drilling unit, each consisting of a water outlet, a monitor and/or hoses and nozzles.
- A monitor allows a single person to control high-gallonage water flow at a safe distance from a fire.

- The size of a hose determines the amount of water that can flow through it, and therefore how many people it takes to handle it.
- Nozzles can produce either a narrow stream of water or wide fog of droplets.
- Foam systems use the fire-main system as a water source and a special nozzle and a proportioner to mix in foam concentrate and agitate the mixture to form a foam for extinguishing burning liquids.
- Enclosed spaces use a system of CO_2 cylinders and piping to deliver carbon dioxide to smother a fire.
- Automatic sprinklers protect living, recreation, office, and galley areas.
- A water spray system with a deluge valve protects the well-head.

Fire Detectors

- A fire line automatic system uses pressurized tubing that triggers an alarm when a hole burns through it.
- Heat and smoke detectors detect high temperatures or the presence of smoke in an enclosed space.
- Combustible-gas detectors warn of high concentrations of any flammable gas, which can lead to fire.
- Manual fire alarms around the rig are used when a crew member detects a fire.

Personal Protective Clothing

- A bunker suit consists of a helmet, hood, coat and pants or long coat, boots, and an air supply.
- A bunker suit protects the wearer from burns, smoke, dangerous gases, and flying objects and liquids.

Glossary

▼
▼
▼

A

adjustable choke *n*: a choke in which the position of a conical needle, sleeve, or plate may be changed with respect to their seat to vary the rate of flow; may be manual or automatic. See *choke*.

aerosol *n*: suspension of very small particles in a gas.

air motor *n*: a motor powered by compressed air.

alarm *n*: a warning device triggered by the presence of abnormal conditions in a machine or system. For example, a low-water alarm automatically signals when the water level in a vessel falls below its preset minimum. Offshore, alarms are used to warn personnel of dangerous or unusual conditions, such as fires and escaping gases.

analog data *n*: information indicated by a continuous form, usually a needle or pointer moving across a dial face. Compare *digital readout*.

analog signal *n*: the representation of the magnitude of a variable in the form of a measurable physical quantity that varies smoothly rather than in discrete steps.

attack hoses *n pl*: smaller-sized hose used to put out small fires or a fire that is more or less under control.

automatic choke *n*: an adjustable choke that is power-operated to control pressure of flow. See *adjustable choke*.

automatic control *n*: a device that regulates various factors (such as flow rate, pressure, or temperature) of a system without supervision or operation by personnel. See *instrumentation*.

automatic driller *n*: a mechanism used to regulate the amount of weight on the bit without the presence of personnel. Automatic drillers free the driller from the sometimes tedious task of manipulating the drawworks brake to maintain correct weight on the bit. Also called an automatic drilling control unit.

automatic slips *n*: see *power slips*.

automation *n*: automatic, self-regulating control of equipment, systems, or processes. See *instrumentation*.

auxiliaries *n pl*: equipment on a drilling or workover rig that is not a direct part of the rig's drilling equipment, such as the equipment used to generate electricity for rig lighting or the equipment used to mix drilling fluid.

B

bail *n*: the U-shaped handle on a nozzle used to close a valve that shuts off water flow when pushed forward.

blowdown *n*: 1. the emptying or depressurizing of material in a vessel. 2. the material thus discarded.

blowout preventer *n*: one of several valves installed at the wellhead to prevent the escape of pressure either in the annular space between the casing and the drill pipe or in open hole (i.e., hole with no drill pipe) during drilling or completion operations. Blowout preventers on land rigs are located beneath the rig at the land's surface; on jackup or platform rigs, at the water's surface; and on floating offshore rigs, on the seafloor.

boom *n*: a movable arm of tubular or bar steel used on some types of cranes or derricks to support the hoisting lines that carry the load.

BOP *abbr*: blowout preventer.

bottleneck *n*: an area of reduced diameter in pipe caused by excessive longitudinal strain or by a combination of longitudinal strain and the swaging action of a body. A bottleneck may result if the downward motion of the drill pipe is stopped with the slips instead of the brake.

bowl *n*: in drilling operations, an insert that fits into the opening of a master bushing and accommodates the slips. Also called an insert bowl or insert.

brackish water *n*: water that contains relatively low concentrations of soluble salts. Brackish water is saltier than fresh water but not as salty as salt water.

break out *v*: to unscrew one section of pipe from another section, especially drill pipe while it is being withdrawn from the wellbore. During this operation, the tongs are used to start the unscrewing operation.

brine *n*: water that has a large quantity of salt, especially sodium chloride, dissolved in it; salt water.

bunker suit *n*: protective firefighting clothing stored on a rig that consists of a helmet, hood, coat, pants, gloves, boots, and oxygen supply.

C

calibration *n*: the adjustment or standardizing of a measuring instrument or of a standard capacity measure.

carbon tetrachloride *n*: a liquid used for degreasing metals; it is toxic when ingested, breathed, or exposed to the skin and is a known carcinogen.

casing *n*: steel pipe placed in an oil or gas well to prevent the wall of the hole from caving in, to prevent movement of fluids from one formation to another, and to improve the efficiency of extracting petroleum if the well is productive. A joint of casing may be 16 to 48 feet (4.8 to 14.6 metres) long and from 4.5 to 20 inches (11.4 to 50.8 centimetres) in diameter.

casing slip *n*: see *spider*.

casing spider *n*: see *spider*.

centrifugal compressor *n*: a compressor in which the flow of gas to be compressed is moved away from the center rapidly, usually by a series of blades or turbines. It is a continuous-flow compressor with a low-pressure ratio and is used to transmit gas through a pipeline. Gas passing through the compressor contacts a rotating impeller, from which it is discharged into a diffuser, where its velocity is slowed and its kinetic energy changed to static pressure. Centrifugal compressors are nonpositive-displacement machines, often arranged in series on a line to achieve multistage compression.

choke *n*: a device with an orifice installed in a line to restrict the flow of fluids. Surface chokes are part of the Christmas tree on a well and contain a choke nipple, or bean, with a small-diameter bore that serves to restrict the flow. Chokes are also used to control the rate of flow of the drilling mud out of the hole when the well is closed in with the blowout preventer and a kick is being circulated out of the hole.

column racker *n*: an automated crane and rack system that stores pipe vertically and moves it to the well center and back.

condensation *n*: the process by which vapors are converted into liquids, chiefly accomplished by cooling the vapors, lowering the pressure on the vapors, or both.

condenser *n*: a form of heat exchanger in which the heat in vapors is transferred to a flow of cooling water or air, causing the vapors to form a liquid.

crane *n*: a machine for raising, lowering, and revolving heavy pieces of equipment, especially on offshore rigs and platforms.

crew *n*: 1. the workers on a drilling or workover rig, including the driller, the derrickhand, and the rotary helpers. 2. any group of oilfield workers.

cuttings *n pl*: the fragments of rock dislodged by the bit and brought to the surface in the drilling mud. Washed and dried cuttings samples are analyzed by geologists to obtain information about the formations drilled.

D

database *n*: a complete collection of information, such as contained on magnetic disks or in the memory of an electronic computer.

day tank *n*: a fuel tank in the fuel supply system for a diesel engine between the main supply tank and the engine that holds a limited amount of fuel.

debug *v*: to detect, locate, and correct malfunctions in a computer, instrumentation, or other type of system.

defensive hoses *n pl*: large-size hoses used to throw large amounts of water from a distance.

deluge valve *n*: valve that opens quickly and allows a full volume of water to flow immediately.

demister *n*: in an evaporator, a separator that removes droplets of liquid from water vapor.

diesel fuel *n*: a light hydrocarbon mixture for diesel engines, similar to furnace fuel oil; it has a boiling range just above that of kerosene.

digital *adj*: pertaining to data in the form of digits, especially electronic data stored in the form of a binary code.

digital readout *n*: a type of register on which the information is indicated by directly readable characters, particularly numerals. Compare *analog data*.

digital signal *n*: the representation of the magnitude of a variable in the form of discrete values or pulses of a measurable physical quantity.

directional drilling *n*: intentional deviation of a wellbore from the vertical. Although wellbores are normally drilled vertically, it is sometimes necessary or advantageous to drill at an angle from the vertical. Controlled directional drilling makes it possible to reach subsurface areas laterally remote from the point where the bit enters the earth. It often involves the use of deflection tools.

directional hole *n*: a wellbore intentionally drilled at an angle from the vertical. See *directional drilling*.

distillate *n*: a product of distillation, i.e., the liquid condensed from the vapor produced in an evaporator. Sometimes called condensate.

distillation *n*: the process of driving off gas or vapor from liquids or solids, usually by heating, and condensing the vapor back to liquid to purify, fractionate, or form new products.

dope *n*: a lubricant for the threads of oilfield tubular goods. *v:* to apply thread lubricant.

doping *n*: applying lubricant (dope) to a tool joint or other threaded connection.

downhole *adj*, *adv*: pertaining to the wellbore.

driller *n*: the employee directly in charge of a drilling or workover rig and crew. The driller's main duty is operation of the drilling and hoisting equipment, but this person is also responsible for downhole condition of the well, operation of downhole tools, and pipe measurements.

driller's console *n*: a metal cabinet on the rig floor containing the controls that the driller manipulates to operate various components of the drilling rig.

driller's control panel *n*: see *driller's console*.

drilling mud *n*: a specially compounded liquid circulated through the wellbore during rotary drilling operations. Also called drilling fluid, mud.

drilling parameters *n pl*: factors that affect a drilling operation, such as the rate of penetration, pump rate, rotary revolutions per minute (rpm), and weight on bit.

drill pipe *n*: seamless steel or aluminum pipe made up in the drill stem between the kelly or top drive on the surface and the drill collars on the bottom. During drilling, it is usually rotated while drilling fluid is circulated through it. Joints of pipe about 30 feet (9 metres) long are coupled together by means of tool joints.

E

electric rig *n*: a drilling rig on which the energy from the power source—usually several diesel engines—is changed to electricity by generators mounted on the engines. The electrical power is then distributed through electrical conductors to electric motors. The motors power the various rig components. Compare *mechanical rig*.

engine *n*: a machine for converting the heat content of fuel into rotary motion that can be used to power other machines. Compare *motor*.

evaporator *n*: a vessel used to convert a liquid into its vapor phase.

exchanger *n*: a piping arrangement that permits heat from one fluid to be transferred to another fluid as they travel countercurrently to one another.

explosion-proof motor *n*: a motor with an enclosure designed to contain an internal explosion and to prevent ignition of surrounding gases or vapors by sparks that may occur in the motor.

F

fatigue *n*: the tendency of material such as a metal to break under repeated cyclic loading at a stress considerably less than the tensile strength shown in a static test.

filter *n*: a porous medium through which a fluid is passed to separate particles of suspended solids from it.

fingerboard *n*: a rack that supports the tops of the stands of pipe being stacked in the derrick or mast. It has several steel fingerlike projections that form a series of slots into which the derrickman can place a stand of drill pipe after it is pulled out of the hole and removed from the drill string.

fog *n*: a wide, fine spray from a nozzle.

formation *n*: a bed or deposit composed throughout of substantially the same kind of rock; often a lithologic unit. Each formation is given a name, frequently as a result of the study of the formation outcrop at the surface and sometimes based on fossils found in the formation.

G

gallonage *n*: the amount of liquid a fire-fighting device delivers in U.S. gallons.

gauge *n*: a device (such as a pressure gauge) used to measure some physical property.

grease fitting *n*: a device on a machine that is designed to accept the hose of a grease gun so that grease can be added to the part in need of lubrication.

H

heat exchanger *n*: see *exchanger*.

hook load *n*: the weight of a drill stem that is suspended from the hook.

horizontal drilling *n*: deviation of the borehole at least 80° from vertical so that the borehole penetrates a productive formation in a manner parallel to the formation. A single horizontal hole can effectively drain a reservoir and eliminate the need for several vertical boreholes.

hydraulic *adj*: 1. of or relating to water or other liquid in motion. 2. operated, moved, or effected by water or liquid.

hydraulic fluid *n*: a liquid of low viscosity (such as light oil) that is used in systems actuated by liquid (such as the brake system in a modern passenger car).

hydraulic force *n*: force resulting from pressure on water or other hydraulic fluid.

I

ignition temperature *n*: the temperature at which a particular vapor burns.

in-line proportioner *n*: a venturi tube that draws foam concentrate into a water stream at a predetermined proportion.

insert *n*: a removable part molded to be set into the opening of the master bushing so that various sizes of slips may be accommodated. Also called a bowl or insert bowl.

instrumentation *n*: a device or assembly of devices designed for one or more of the following functions: to measure operating variables (such as pressure, temperature, rate of flow, and speed of rotation); to indicate these phenomena with visible or audible signals; to record them; to control them within a predetermined range; and to stop operations if the control fails. Simple instrumentation might consist of an indicating pressure gauge only. In a completely automatic system, desired ranges of pressure, temperature, and so on are predetermined and preset.

Iron Roughneck™ *n*: a manufacturer's name for a floor-mounted combination of a spinning wrench and a torque wrench. The Iron Roughneck moves into position hydraulically and eliminates the manual handling involved with suspended individual tools.

J

joint *n*: in drilling, a single length (from 16 feet to 45 feet, or 5 metres to 14.5 metres, depending on its range length) of drill pipe, drill collar, casing, or tubing that has threaded connections at both ends. Several joints screwed together constitute a stand of pipe.

K

kelly *n*: the heavy steel tubular device, four- or six-sided, suspended from the swivel through the rotary table and connected to the top joint of drill pipe to turn the drill stem as the rotary table turns. It has a bored passageway that permits fluid to be circulated into the drill stem and up the annulus, or vice versa. Kellys manufactured to API specifications are available only in four- or six-sided versions, are either 40 or 54 feet (12 to 16 metres) long, and have diameters as small as 2½ inches (6 centimetres) and as large as 6 inches (15 centimetres).

kelly spinner *n*: a pneumatically operated device mounted on top of the kelly that, when actuated, causes the kelly to turn or spin. It is useful when the kelly or a joint of pipe attached to it must be spun up, that is, rotated rapidly for being made up.

kick *n*: an entry of water, gas, oil, or other formation fluid into the wellbore during drilling. It occurs because the pressure exerted by the column of drilling fluid is not great enough to overcome the pressure exerted by the fluids in the formation drilled. If prompt action is not taken to control the kick, or kill the well, a blowout may occur.

L

landing string *n*: offshore, to land the wellhead on casing beneath the water on the seafloor, it is run in a landing string (drill pipe or tubing) to seafloor or inside casing.

LCD *abbr*: liquid crystal display.

liquefied petroleum gas (LPG) *n*: a mixture of heavier, gaseous, paraffinic hydrocarbons, principally butane and propane. These gases, easily liquefied at moderate pressure, may be transported as liquids and converted to gases on release of the pressure. Thus, liquefied petroleum gas is a portable source of thermal energy that finds wide application in areas where it is impractical to distribute natural gas. It is also used as a fuel for internal-combustion engines and has many industrial and domestic uses. Principal sources are natural and refinery gas, from which the liquefied petroleum gases are separated by fractionation.

liquid crystal display (LCD) *n*: a readout that shows information in the form of lighted marks against a dark background.

load *n*: in mechanics, the weight or pressure placed on an object. The load on a bit refers to the amount of weight of the drill collars allowed to rest on the bit. See *weight on bit*.

logging while drilling (LWD) *n*: logging measurements obtained by measurement-while-drilling techniques as the well is being drilled.

LPG *abbr*: liquefied petroleum gas.

lubricate *v*: 1. to apply grease or oil to moving parts. 2. to lower or raise tools in or out of a well with pressure inside the well. The term comes from the fact that a lubricant (grease) is often used to provide a seal against well pressure while allowing wireline to move in or out of the well.

LWD *abbr*: logging while drilling.

M

make up *v*: 1. to assemble and join parts to form a complete unit (e.g., to make up a string of drill pipe). 2. to screw together two threaded pieces. Compare *break out*.

measurement while drilling (MWD) *n*: 1. directional and other surveying during routine drilling operations to determine the angle and direction by which the wellbore deviates from the vertical. 2. any system of measuring downhole conditions during routine drilling operations.

measurement-while-drilling system *n*: a system in which downhole conditions are monitored during the drilling of a well.

mechanical nozzle *n*: used in a foam system where the venturi device and the agitation device are in the same nozzle.

mechanical rig *n*: a drilling rig in which the source of power is one or more internal-combustion engines and in which the power is distributed to rig components through mechanical devices (such as chains, sprockets, clutches, and shafts). Also called a power rig. Compare *electric rig*.

microprocessor *n*: a circuit card that interprets information.

monkeyboard *n*: the derrickhand's working platform. As pipe or tubing is run into or out of the hole, the derrickhand must handle the top end of the pipe, which may be as high as 90 feet (27 metres) or higher in the derrick or mast. The monkeyboard provides a small platform to raise the derrickhand to the proper height for handling the top of the pipe.

motor *n*: any of various power units, such as a hydraulic, air, or electric device, that develops energy or imparts motion. Compare *engine*.

mousehole *n*: an opening in the rig floor, usually lined with pipe, into which a length of drill pipe is placed temporarily for later connection to the drill string.

mousehole connection *n*: the procedure of adding a length of drill pipe or tubing to the active string. The length to be added is placed in the mousehole, made up to the kelly, then pulled out of the mousehole and subsequently made up into the string. Compare *rathole connection*.

MWD *abbr*: measurement while drilling.

MWD directional survey *n*: a directional survey that uses measurement-while-drilling techniques to determine drift angle and azimuth.

N

nozzle *n*: a narrow passageway at the end of a hose or tube that causes a fluid to be ejected at high velocity. Different-size nozzles produce different-velocity sprays.

O

offshore *n*: that geographic area that lies seaward of the coastline. In general, the term "coastline" means the line of ordinary low water along that portion of the coast that is in direct contact with the open sea or the line marking the seaward limit of inland waters.

offshore drilling *n*: drilling for oil or gas in an ocean, gulf, or sea, usually on the Outer Continental Shelf. A drilling unit for offshore operations may be a mobile floating vessel with a ship or barge hull, a semisubmersible or submersible base, a self-propelled or towed structure with jacking legs (jackup drilling rig), or a permanent structure used as a production platform when drilling is completed. In general, wildcat wells are drilled from mobile floating vessels or from jackups, while development wells are drilled from platforms or jackups.

oil-base mud *n*: a drilling or workover fluid in which oil is the continuous phase and which contains from less than 2 percent and up to 5 percent water. This water is spread out, or dispersed, in the oil as small droplets.

operator *n*: the person or company, either proprietor or lessee, actually operating an oilwell or lease, generally the oil company that engages the drilling, service, and workover contractors.

osmosis *n*: passage of a pure liquid, such as water, into a solution, such as salt water, through a membrane that allows the water to pass through but not the salt. See *reverse osmosis*.

osmotic pressure *n*: the force that moves a liquid through a barrier; it is powerful enough to raise sap from the roots to the tops of trees.

oxidation *n*: 1. the process of burning. 2. a chemical reaction with oxygen in which a compound loses electrons and gains a more positive charge.

P

permeability *n*: 1. a measure of the ease with which molecules flow through connecting pore spaces of rock or cement. The unit of measurement is the millidarcy. 2. fluid conductivity of a porous medium. 3. ability of a fluid to flow within the interconnected pore network of a porous medium.

permeator *n*: the chamber in a reverse osmosis watermaker where reverse osmosis takes place.

pickup-and-laydown system *n*: a boom that lifts pipe from a horizontal position at the V-door and tilts it to vertical and moves it over the mousehole or well center.

pipe *n*: a long, hollow cylinder, usually steel, through which fluids are conducted. Oilfield tubular goods are casing (including liners), drill pipe, tubing, or line pipe.

pipe rack *n*: a horizontal support for tubular goods.

pneumatic *adj*: operated by air pressure.

pneumatic control *n*: a control valve that is actuated by air. Several pneumatic controls are used on drilling rigs to actuate and control rig components (such as clutches, hoists, engines, and pumps).

pneumatic line *n*: any hose or line, usually reinforced with steel, that conducts air from an air source (such as a compressor) to a component that is actuated by air (such as a clutch).

pneumatic tube fire detector *n*: a system that can detect fires in open spaces, consisting of a length of flexible plastic or metal tubing that forms a loop around the outside of the structure it protects.

power slips *n pl*: a device, operated by air or hydraulic fluid, that fits into the opening in the rotary table when the drill stem must be suspended in the wellbore (as when a connection or trip is being made). Power slips, also called automatic slips, eliminate the need for the crew to set and take out slips manually. See *slips*.

power tools *n pl*: equipment operated hydraulically or by compressed air for making up and breaking out drill pipe, casing, tubing, rods, nuts, and so on.

pump *n*: a device that increases the pressure on a fluid or raises it to a higher level.

pump rate *n*: the speed, or velocity, at which a pump is run. In drilling, the pump rate is usually measured in strokes per minute (spm).

R

rack pipe *v*: 1. to place pipe withdrawn from the hole on a pipe rack. 2. to stand pipe on the derrick floor when pulling it out of the hole.

radiant heat *n*: the heat from a fire that travels in all directions.

rathole *n*: a hole in the rig floor, some 30 to 40 feet (9 to 12 metres) deep, which is lined with casing that projects above the floor, into which the kelly and the swivel are placed when hoisting operations are in progress.

rathole connection *n*: the addition of a length of drill pipe or tubing to the active string using the rathole instead of the mousehole, which is the more common connection. The length to be added is placed in the rathole, made up to the kelly, pulled out of the rathole, and made up into the string. Compare *mousehole connection*.

readout *n*: a device that displays numbers or symbols and incorporates electric or electronic features.

recording gauge *n*: a device, driven by a clockwork mechanism, that provides a chronological record of gauge indications (e.g., by tracing values of pressure, vacuum, voltage) on a paper form.

recording instrument *n*: a measuring instrument that records the value of the measured variable by marking or printing on a removable paper chart, tape, or other suitable recording material.

reverse osmosis *n*: a method of desalting brackish or salt water by passing it through a membrane that is not permeable to salt. See *osmosis*.

rig floor *n*: the area immediately around the rotary table and extending to each corner of the derrick or mast—that is, the area immediately above the substructure on which the drawworks, the rotary table, and so forth rest. Also called derrick floor, drill floor.

rotary speed *n*: the speed, measured in revolutions per minute, at which the rotary table is operated.

rotary table *n*: the principal piece of equipment in the rotary table assembly; a turning device used to impart rotational power to the drill stem while permitting vertical movement of the pipe for rotary drilling. The master bushing fits inside the opening of the rotary table; it turns the kelly bushing, which permits vertical movement of the kelly while the stem is turning.

rotary torque *n*: the rotational force applied to turn the drill stem.

rpm *abbr*: revolutions per minute.

S

salt water *n*: a water that contains a large quantity of salt, i.e., brine.

scale *n*: a mineral deposit (e.g., calcium carbonate) that precipitates out of water and adheres to the inside of pipes, heaters, and other equipment.

semipermeable *n*: the ability to let small molecules pass or flow through.

slips *n pl*: wedge-shaped pieces of metal with serrated inserts (bowls) or other gripping elements, such as serrated buttons, that suspend the drill pipe or drill collars in the master bushing of the rotary table when it is necessary to disconnect the drill stem from the kelly or from the top-drive unit's drive shaft. Rotary slips fit around the drill pipe and wedge against the master bushing to support the pipe. Drill collar slips fit around a drill collar and wedge against the master bushing to support the drill collar. Power slips are pneumatically or hydraulically actuated devices that allow the crew to dispense with the manual handling of slips when making a connection.

slip segment *n*: a single part of all the parts, or segments, that make up the slips. See *slips*.

solvent *n*: a substance, usually liquid, in which another substance (the solute) dissolves.

spider *n*: a circular steel device that holds slips supporting a suspended string of drill pipe, casing, or tubing. A spider may be split or solid.

spm *abbr*: strokes per minute.

spring slips *n pl*: slips that set by means of springs when a floorhand stands on them. See *slips*.

steam *n*: water in its gaseous state.

T

top drive *n*: a device similar to a power swivel that is used in place of the rotary table to turn the drill stem. It also includes power tongs. Modern top drives combine the elevator, the tongs, the swivel, and the hook. Even though the rotary table assembly is not used to rotate the drill stem and bit, the top-drive system retains it to provide a place to set the slips to suspend the drill stem when drilling stops.

torque *n*: the turning force that is applied to a shaft or other rotary mechanism to cause it to rotate or tend to do so. Torque is measured in foot-pounds, joules, newton-metres, and so forth.

trip *n*: the operation of hoisting the drill stem from and returning it to the wellbore. *v:* shortened form of "make a trip."

trip in *v*: to go in the hole.

trip out *v*: to come out of the hole.

V

vacuum *n*: 1. a space that is theoretically devoid of all matter and that exerts zero pressure. 2. a condition that exists in a system when pressure is reduced below atmospheric pressure.

valve *n*: a device used to control the rate of flow in a line to open or shut off a line completely, or to serve as an automatic or semiautomatic safety device.

vapor *n*: a substance in the gaseous state that can be liquefied by compression or cooling.

vaporization *n*: 1. the act or process of converting a substance into the vapor phase. 2. the state of substances in the vapor phase.

V-door *n*: an opening at floor level in a side of a derrick or mast. The V-door is opposite the drawworks and is used as an entry to bring in drill pipe, casing, and other tools from the pipe rack. The name comes from the fact that on the old standard derrick, the shape of the opening was an inverted V.

venturi effect *n*: the drop in pressure resulting from the increased velocity of a fluid as it flows through a constricted section of a pipe.

venturi tube *n*: a short tube with a calibrated constriction that is used in instruments or devices such as jet hoppers. It was developed to take advantage of the principle that a fluid flowing through a constriction has increased velocity and reduced pressure.

W

watermaker *n*: a piece of equipment that purifies brackish or salt water into fresh water.

weight indicator *n*: an instrument near the driller's position on a drilling rig that shows both the weight of the drill stem that is hanging from the hook (hook load) and the weight that is placed on the bit by the drill collars (weight on bit).

weight on bit (WOB) *n*: the amount of downward force placed on the bit by the weight of the drill collars.